Statistics and Chemometrics for Analytical Chemistry

Fourth Edition

Pearson
Education

Statistics and Chemometrics for Analytical Chemistry

Fourth Edition

James N. Miller and Jane C. Miller

An imprint of Pearson Education

Harlow, England · London · New York · Reading, Massachusetts · San Francisco · Toronto · Don Mills, Ontario · Sydney
Tokyo · Singapore · Hong Kong · Seoul · Taipei · Cape Town · Madrid · Mexico City · Amsterdam · Munich · Paris · Milan

Pearson Education Limited
Edinburgh Gate
Harlow
Essex CM20 2JE
England

and Associated Companies throughout the world

Visit us on the World Wide Web at:
www.pearsoneduc.com

Third edition published under the Ellis Horwood imprint 1993
Fourth edition 2000

ISBN 0 130 22888 5

British Library Cataloguing-in-Publication Data
A catalogue record for this book can be obtained from the British Library

Library of Congress Cataloging-in-Publication Data
A catalog record for this book can be obtained from the Library of Congress

10 9 8 7 6 5 4 3 2
05 04 03 02 01

Typeset by 60
Printed in Great Britain by Henry Ling Ltd., at the Dorset Press,
Dorchester, Dorset

Contents

Preface to the first edition

To add yet another volume to the already numerous texts on statistics might seem to be an unwarranted exercise, yet the fact remains that many highly competent scientists are woefully ignorant of even the most elementary statistical methods. It is even more astonishing that analytical chemists, who practise one of the most quantitative of all sciences, are no more immune than others to this dangerous, but entirely curable, affliction. It is hoped, therefore, that this book will benefit analytical scientists who wish to design and conduct their experiments properly, and extract as much information from the results as they legitimately can. It is intended to be of value to the rapidly growing number of students specializing in analytical chemistry, and to those who use analytical methods routinely in everyday laboratory work.

There are two further and related reasons that have encouraged us to write this book. One is the enormous impact of microelectronics, in the form of microcomputers and hand-held calculators, on statistics: these devices have brought lengthy or difficult statistical procedures within the reach of all practising scientists. The second is the rapid development of new 'chemometric' procedures, including pattern recognition, optimization, numerical filter techniques, simulations and so on, all of them made practicable by improved computing facilities. The last chapter of this book attempts to give the reader at least a flavour of the potential of some of these newer statistical methods. We have not, however, included any computer programs in the book – partly because of the difficulties of presenting programs that would run on all the popular types of microcomputer, and partly because there is a substantial range of suitable and commercially available books and software.

The availability of this tremendous computing power naturally makes it all the more important that the scientist applies statistical methods rationally and correctly. To limit the length of the book, and to emphasize its practical bias, we have made no attempt to describe in detail the theoretical background of the statistical tests described. But we have tried to make it clear to the practising analyst which tests are appropriate to the types of problem likely to be encountered in the laboratory. There are worked examples in the text, and exercises for the reader at the end of each chapter. Many of these are based on the data provided by research papers published in *The Analyst.* We are deeply grateful to Mr. Phil Weston, the Editor, for allowing us thus to make use of his distinguished journal. We also thank our colleagues, friends and family for their forbearance during the preparation of the book; the sources of the statistical tables, individually acknowledged in the appendices; the Series Editor, Dr. Bob Chalmers; and our publishers for their efficient cooperation and advice.

J. C. Miller
J. N. Miller
April 1984

Preface to the fourth edition

Since the publication of the third edition of this book in 1993 the use of elementary and advanced statistical methods in the teaching and the practice of the analytical sciences has continued to grow rapidly. This new edition attempts to keep pace with these developments while retaining the basic approach of previous editions, which adopted a pragmatic and as far as possible non-mathematical approach to statistical calculations.

A major change in the last few years has been the much wider use of the more advanced methods of multivariate analysis. This has been reflected by the addition of an extra chapter to the book (Chapter 8) which provides a more detailed introduction to such methods, without venturing into the realms of the matrix algebra that underpins them. We have also felt encouraged to alter the title of the book to reflect the wider use of these chemometric techniques. This term is now sometimes used to cover both elementary and multivariate statistical methods as applied to chemistry, but we have retained the view that it should be applied to the more advanced calculations that require the power of personal computers.

All students, researchers and laboratory personnel now have access to such computers. As usual, the availability of such a range of statistical techniques enhances rather than diminishes the need for a full understanding of these methods. It has also encouraged us to include in the text examples of calculations performed by two established pieces of software, Excel and Minitab. The former is probably accessible from the majority of desktops, and is widely used in the collection and processing of data from analytical instruments, while the latter is frequently adopted in education as well as by practising scientists. In each program the calculations, at least the simple ones used in this book, are easily accessible and simply displayed, and many texts are available as general introductions to the software. Moreover, additional facilities such as graphical displays, regression diagnostics, etc. are available from these programs, providing opportunities for better understanding and further data interpretation. These extra facilities are utilized in some examples provided in the Instructors' Manual, which accompanies this edition of our book for the first time. The Manual also contains ideas for classroom and laboratory work, a complete set of figures for use as OHP masters, and fully worked solutions to the exercises in this volume: this text now contains only outline solutions.

Another area where rapid strides have been made in the analytical sciences has been the issue of the quality of analytical results. For this reason Chapter 4, which covers a range of relevant topics, has been substantially rewritten and expanded for this edition. The treatment of control charts has been extended, and the important areas of proficiency testing schemes and collaborative trials have been given much fuller coverage. Other areas of statistics where we have tried to give more details are several aspects of regression and calibration, robust methods and the treatment of outliers, exploratory data analysis, analysis of variance, and experimental design and optimization. Appendix 1 has been re-cast to try to offer more guidance on the vexed question of the most suitable statistical test to be used in a particular situation, and the inclusion of more significance tests has meant that the statistical tables have also been expanded. A few topics covered in earlier editions but with limited practical application have been omitted to make space for these new areas.

We are very grateful to many correspondents and staff and student colleagues who continue to provide us with constructive comments and suggestions, and to point out minor errors and omissions. We also thank the Royal Society of Chemistry for permission to use data from papers published in *The Analyst*. Finally we thank Alex Seabrook and her editorial colleagues at Pearson Education, for their perfect mixture of expertise, patience and enthusiasm.

James N. Miller
Jane C. Miller
October 1999

Acknowledgements

The publishers wish to thank the following for permission to reproduce copyright material:

Tables A.2, A.3, A.4, A.7, A.8, A.11, A.12, A.13 and A.14 reproduced with the permission of Routledge; Table A.5 reprinted with permission from the *Journal of the American Statistical Association*, copyright 1958 by the American Statistical Association. All rights reserved; Table A.6 reproduced by permission of John Wiley & Sons, Limited; Table A.10 adapted with the permission of the Institute of Mathematical Statistics; Data from articles published in *The Analyst* used with the permission of the Royal Society of Chemistry; Examples of Minitab input and output used with the permission of Minitab Inc.

A Companion Web Site accompanies *Statistics and Chemometrics for Analytical Chemistry* by Miller and Miller

Visit the ***Statistics and Chemometrics for Analytical Chemistry*** Companion Web Site at www.booksites.net/miller/ to find valuable teaching and learning material including:

For students:

- Study material designed to help you improve your results
- Links to valuable resources on the web
- Selected charts from the book in downloadable form

For lecturers:

- A secure, password protected site with teaching material
- A syllabus manager that will build and host a course web page
- An electronic, downloadable version of the Instructor's Manual, including complete worked answers to questions in the book

Minitab Inc. can be contacted at the following address:

Minitab Inc.
3081 Enterprise Drive
State College, PA 16801 USA
Telephone: 814 238 3280
Fax: 814 238 4383
e-mail: Info@minitab.com
URL: http://www.minitab.com

Glossary of symbols

a	–	intercept of regression line
b	–	gradient of regression line
c	–	number of columns in two-way ANOVA
C	–	correction term in two-way ANOVA
C	–	used in Cochran's text for homogeneity of variance
F	–	the ratio of two variances
G	–	used in Grubbs' test for outliers
h	–	number of samples in one-way ANOVA
μ	–	arithmetic mean of a population
M	–	number of minus signs in Wald–Wolfowitz runs test
n	–	sample size
N	–	number of plus signs in Wald–Wolfowitz runs test
N	–	total number of measurements in two-way ANOVA
ν	–	number of degrees of freedom
$P(r)$	–	probability of r
Q	–	Dixon's Q, used to test for outliers
r	–	product–moment correlation coefficient
r	–	number of rows in two-way ANOVA
r	–	number of smallest and largest observations omitted in trimmed mean calculations
R^2	–	coefficient of determination
R'^2	–	adjusted coefficient of determination
r_s	–	Spearman rank correlation coefficient
s	–	standard deviation of a sample
$s_{y/x}$	–	standard deviation of y-residuals
s_b	–	standard deviation of slope of regression line
s_a	–	standard deviation of intercept of regression line
$s_{(y/x)_w}$	–	standard deviation of y-residuals of weighted regression line
s_{x_0}	–	standard deviation of x-value estimated using regression line

s_B – standard deviation of blank
s_{x_E} – standard deviation of extrapolated x-value
$s_{x_{0W}}$ – standard deviation of x-value estimated by using weighted regression line
σ – standard deviation of a population
σ_0^2 – measurement variance
σ_1^2 – sampling variance
t – quantity used in the calculation of confidence limits and in significance testing of mean (see Section 2.4)
T – grand total in ANOVA
T_1 and T_2 – test statistics used in the Wilcoxon rank sum test
w – range
w_i – weight given to point on regression line
$\bar{x}$ – arithmetic mean of a sample
x_0 – x-value estimated by using regression line
x_0 – outlier value of x
$\tilde{x}_i$ – pseudo-value in robust statistics
x_E – extrapolated x-value
$\bar{x}_W$ – arithmetic mean of weighted x-values
X^2 – quantity used to test for goodness-of-fit
$\hat{y}$ – y-values predicted by regression line
$\bar{y}_W$ – arithmetic mean of weighted y-values
y_B – signal from blank
z – standard normal variable

CHAPTER ONE

Introduction

1.1 Analytical problems

An analytical chemist may be presented with two types of problem. Sometimes only a qualitative answer is required. For example, the presence of boron in distilled water is very damaging in the manufacture of microelectronic components – 'Does this distilled water sample contain any boron?' Again, the comparison of soil samples is a common problem in forensic science – 'Could these two soil samples have come from the same site?' In other cases the problems posed are quantitative ones. 'How much albumin is there in this sample of blood serum?' 'How much lead in this sample of tap-water?' 'This steel sample contains small quantities of chromium, tungsten and manganese – how much of each?' These are typical examples of single-component or multiple-component quantitative analyses.

Modern analytical chemistry is overwhelmingly a **quantitative** science. In many cases a quantitative answer will be much more valuable than a qualitative one. It may be useful for an analyst to claim to have detected some boron in a distilled water sample, but it is much more useful to be able to say *how much* boron is present. The person who requested the analysis could, armed with this quantitative answer, judge whether or not the level of boron was of concern, consider how it might be reduced, and so on. But if it was known only that *some* boron was present it would be hard to judge the significance of the result. In other cases, it is only a quantitative result that has any value at all. For example, almost all samples of (human) blood serum contain albumin; the only question is, how much?

It is important to note that even where a qualitative answer is required, quantitative methods are used to obtain it. This point is illustrated with the aid of the examples given at the beginning of this section. In reality, an analyst would never simply report 'I can/cannot detect boron in this water sample'. A quantitative method capable of detecting boron at, say, levels of $1\,\mu g\,ml^{-1}$ would be used. If the test gives a negative result, it would then be described in the form 'This sample contains less than $1\,\mu g\,ml^{-1}$ boron'. If the test gives a positive result the sample will be reported to contain at least $1\,\mu g\,ml^{-1}$ boron (with other information too – see below). Much more complex quantitative approaches might be used to compare two soil samples. For example, the samples might be subjected to a particle size analysis, in which the proportions of the soil particles falling within a number, say 10, of particle-size ranges are determined. Each sample would then be characterized by these 10 pieces of data. Quite complex procedures (see Chapter 8) can then be used to provide a quantitative assessment of their similarity.

1.2 Errors in quantitative analysis

Once we accept that quantitative studies will play a dominant role in any analytical laboratory, we must accept also that the errors that occur in such studies are of supreme importance. Our guiding principle will be that *no quantitative results are of any value unless they are accompanied by some estimate of the errors inherent in them.* This principle naturally applies not only to analytical chemistry but to any field of study in which numerical experimental results are obtained. We can examine a number of simple examples, which not only illustrate the principle but also foreshadow the types of statistical problem we shall meet and solve in subsequent chapters.

Suppose a chemist synthesizes an analytical reagent which he believes to be entirely new. The compound is studied using a spectrometric method and gives a value of 104 (normally, most of our results will be cited in carefully-chosen units, but in this hypothetical example purely arbitrary units can be used). On checking the reference books, the chemist finds that no compound previously discovered has yielded a value of more than 100 when studied by the same method under the same experimental conditions. The question thus naturally arises, has our chemist really discovered a new compound? The answer to this question lies of course in the degree of reliance that we can place on that experimental value of 104. What errors are associated with it? If further study indicates that the result is correct to within 2 (arbitrary) units, i.e. the true value probably lies in the range 104 ± 2, then a new material has probably been discovered. If, however, investigations show that the error may amount to 10 units (i.e. 104 ± 10), then it is quite likely that the true value is actually less than 100, in which case a new discovery is far from certain. In other words, a knowledge of the experimental errors is crucial (in this case as in every other) to the proper interpretation of the results. In statistical terms this example would involve the comparison of the experimental result with an assumed or reference value: this topic is studied in detail in Chapter 3.

A more common situation is that of the analyst who performs several replicate determinations in the course of a single analysis. (The value and significance of such replicates is discussed in detail in the next chapter.) Suppose an analyst performs a titrimetric experiment four times and obtains values of 24.69, 24.73, 24.77 and 25.39 ml. The first point to notice is that titration values are reported to the nearest 0.01 ml; this point is also discussed further in Chapter 2. It is also apparent that all four values are different, because of the errors inherent in the measurements, and that the fourth value (25.39 ml) is substantially different from the other three. So can this fourth value be safely rejected, so that (for example) the mean titre is reported as 24.73 ml, the average of the other three readings? In statistical terms, is the value 25.39 ml an 'outlier'? The important topic of outlier rejection is discussed in detail in Chapters 3 and 6.

Another frequent problem involves the comparison of two (or more) sets of results. Suppose that an analyst measures the vanadium content of a steel sample by two separate methods. With the first method the average value obtained is 1.04%, with an estimated error of 0.07%, and with the second method, the average value is 0.95% with an error of 0.04%. Several questions arise from the comparison of these results. Are the two average values significantly different, or are they indistinguishable within the limits of the experimental errors? Is one method significantly less error-prone than the other? Which of the mean values is actually closer to the truth? Again, Chapter 3 discusses these and related questions.

To conclude this section we note that many analyses are based on graphical methods. Instead of making repeated measurements on the same sample, we perform a series of measurements on a small group of standards, which have known concentrations covering a considerable range. In this way we set up a calibration curve that can be used to estimate the concentrations of test samples studied by the same procedure. In practice, of course, all the measurements (those utilizing the standards and those on the test samples) will be subject to errors. It is necessary, for example, to assess the errors involved in drawing the calibration curve; to estimate the error in the concentration of a single sample determined by using the curve; and to estimate the limit of detection of the method, i.e. the smallest quantity of analyte that can be detected with a particular degree of confidence. These procedures, which are especially commonplace in instrumental analysis, are described in Chapter 5.

These examples represent only a fraction of the possible problems arising from the occurrence of experimental errors in quantitative analysis. As we have seen, however, the problems have to be solved if the quantitative data are to have any real meaning. It is thus apparent that we must study the various types of error in more detail.

1.3 Types of error

Experimental scientists make a fundamental distinction between three types of error. These are known as **gross**, **random**, and **systematic** errors. Gross errors are readily described: they may be defined as errors that are so

Table 1.1 Random and systematic errors

Student			*Results (ml)*			*Comment*
A	10.08	10.11	10.09	10.10	10.12	Precise, biased
B	9.88	10.14	10.02	9.80	10.21	Imprecise, unbiased
C	10.19	9.79	9.69	10.05	9.78	Imprecise, biased
D	10.04	9.98	10.02	9.97	10.04	Precise, unbiased

serious that there is no real alternative to abandoning the experiment and making a completely fresh start. Examples would include a complete instrument breakdown, accidentally dropping or discarding a crucial sample, or discovering during the course of the experiment that a supposedly pure reagent was in fact badly contaminated. Such errors (which occur occasionally even in the best-regulated laboratories!) are normally very easily recognized. In our discussion we thus have only to distinguish carefully between **random** and **systematic** errors.

We can best make this distinction by careful study of a real experimental situation. Four students (A–D) each perform an analysis in which *exactly* 10.00 ml of *exactly* 0.1 M sodium hydroxide is titrated with *exactly* 0.1 M hydrochloric acid. Each student performs five replicate titrations, with the results shown in Table 1.1.

The results obtained by student A have two characteristics. First, they are all very close to each other; all the results lie between 10.08 and 10.12 ml. In everyday terms we would say that the results are highly *reproducible*. The second feature is that they are *all too high*: in this experiment (somewhat unusually) we know the correct answer: the result should be exactly 10.00 ml. It is evident that two entirely separate types of error have occurred within this student's experiment. First, there are **random errors** – *these cause replicate results to differ from one another, so that the individual results fall on both sides of the average value* (10.10 ml in this case). Random errors affect the **precision**, or **reproducibility**, of an experiment. In the case of student A it is clear that the random errors are small, so we say that the results are **precise**. In addition, however, there are **systematic errors** – these cause all the results to be in error in the *same sense* (in this case they are all too high). The total systematic error (note that in a given experiment there may be several sources of systematic error, some positive and others negative, see Chapter 2) is called the **bias** of the measurement. In many experiments the random and systematic errors are not only readily distinguishable by inspection of the results, they also have quite distinct origins in terms of experimental technique and equipment. Before we examine the causes of the errors in this experiment, however, we can discuss briefly the results obtained by students B–D. Student B has obtained results in direct contrast to those of student A. The average of the five results (10.01 ml) is very close to the true value, so there is no evidence of bias. The spread of the results is very large, however, indicating poor precision, i.e. substantial random

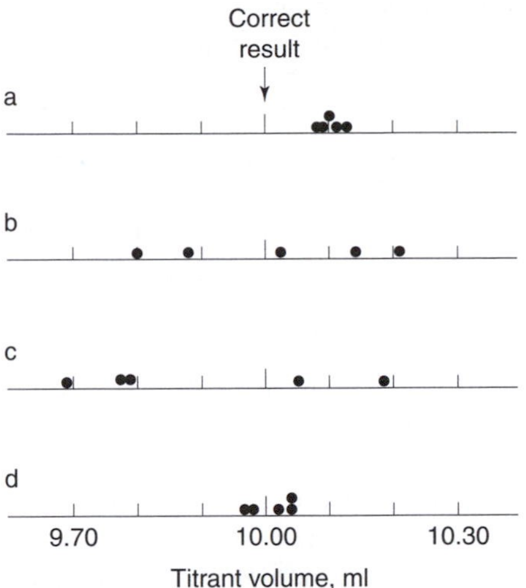

Figure 1.1 Bias and precision – dot-plots of the data in Table 1.1

errors. Comparison of these results with those obtained by student A shows clearly that random and systematic errors can occur independently of one another. This conclusion is reinforced by the data of students C and D. Student C's work has poor precision (range 9.69–10.19 ml) and the average result (9.90 ml) is biased. Student D has achieved both precise (range 9.97–10.04 ml) and unbiased (average 10.01 ml) results. The distinction between random and systematic errors is summarized in Figure 1.1 as a series of dot-plots. This simple graphical method of displaying data, in which individual results are plotted as dots on a linear scale, is frequently used in initial data analysis (see Chapters 3 and 6).

In most analytical experiments the most important question is – how far is the result from the true value of the concentration or amount that we are trying to measure? This is expressed as the **accuracy** of the experiment. Accuracy is defined by the International Standards Organization (ISO) as 'the closeness of agreement between a test result and the accepted reference value' of the analyte. Under this definition the accuracy of a single result may be affected by *both* random and systematic errors. The accuracy of an average result also has contributions from both error sources: even if systematic errors are absent, the average result will probably not equal the reference value exactly, because of the occurrence of random errors (see Chapters 2 and 3). The results obtained by student B exemplify these principles. Four of this student's five measurements show significant inaccuracy, i.e. are well removed from the true value of 10.00. But the average of B's results (10.01) is very accurate, so it seems that the inaccuracy of the individual results is largely due to random errors and not to systematic ones. By contrast, all of student A's individual results, and the resulting average, are inaccurate: given the good precision of this student's work, it seems certain that these inaccuracies are due to systematic errors. It should be noted that, contrary to the implications of many dictionaries, accuracy and precision have entirely different meanings in the study of experimental errors.

In summary, precision describes random error, bias describes systematic error, and the accuracy, i.e. closeness to the true value of a single measurement or a mean value, incorporates both types of error.

Although we earlier used the word 'reproducibility' as an approximate definition of precision, modern convention makes a careful distinction between **reproducibility** and **repeatability**. We can illustrate this distinction by an extension of our previous experiment. In the normal way student A (for example) would do the five replicate titrations in rapid succession; very probably the whole exercise would not take more than an hour or so. The same set of solutions and the same glassware would be used throughout, the same preparation of indicator would be added to each titration flask, and the temperature, humidity and other laboratory conditions would remain much the same. In such circumstances the precision measured would be the *within-run* precision: this is called the **repeatability**. Suppose, however, that for some reason the titrations were performed by different staff on five different occasions in different laboratories, using different pieces of glassware and different batches of indicator. It would not be surprising to find a greater spread of the results in this case. This set of data would reflect the *between-run* precision of the method, i.e. its **reproducibility**.

One further lesson may be learned from the titration experiment. It is easy to appreciate that the data obtained by student C are unacceptable, and that those of student D are the most acceptable. Sometimes, however, it may happen that two methods are available for a particular analysis, one of which is believed to be precise but biased, and the other imprecise but without bias. In other words we may have to choose between the types of results obtained by students A and B respectively. Which type of result is preferable? It is impossible to give a dogmatic answer to this question, if only because in practice the choice of analytical method will often be based on the cost, ease of automation, speed of analysis, and other factors falling outside our elementary evaluation. It is nonetheless important to realize that a method which is substantially free from systematic errors may still, if it is very imprecise, give an average value that is (by chance) a long way from the correct value. On the other hand a method that is precise but biased (e.g. student A) can be converted into one that is both precise *and* unbiased (e.g. student D) if the systematic errors can be discovered and hence removed. There will also be cases where, because the measurements being attempted are completely new, no check for systematic errors is feasible. Random errors can never be eliminated, though by careful technique we can minimize them, and by making repeated measurements we can measure them and evaluate their significance. Systematic errors can in many cases be removed by careful checks on our experimental technique and equipment.

This crucial distinction between the two major types of error is further explored in the next section.

When an analytical laboratory is supplied with a sample and requested to determine the concentrations of one of its constituents, it will doubtless estimate, or perhaps know from experience, the extent of the major random and systematic errors occurring. The customer supplying the sample may well want this information summarized in a single statement, giving the *range within which the true concentration is reasonably likely to lie*. This range, which should be given with a probability (i.e. 'it is 95% probable that the concentration lies between ... and ...'), is called the **uncertainty** of the measurement. This concept, which has its origins in physical metrology, is now rapidly gaining in importance and popularity in analytical chemistry, and is discussed in more detail in Chapter 4.

1.4 Random and systematic errors in titrimetric analysis

The example of the students' titrimetric experiments showed clearly that random and systematic errors can occur independently of one another, and thus presumably arise at different stages of the experiment. Since titrimetry is a relatively simple and still widely adopted procedure it is of interest and value to examine it in detail in this context. A complete titrimetric analysis can be regarded as having the following steps.

1. Making up a standard solution of one of the reactants. This involves (a) weighing a weighing bottle or similar vessel containing some solid material, (b) transferring the solid material to a standard flask and weighing the bottle again to obtain by subtraction the weight of solid transferred (weighing *by difference*), and (c) filling the flask up to the mark with water (assuming that an aqueous titration is to be used).
2. Transferring an aliquot of the standard material to a titration flask with the aid of a pipette. This involves (a) filling the pipette to the appropriate mark, and (b) draining it in a specified manner into the titration flask.
3. Titrating the liquid in the flask with a solution of the other reactant, added from a burette. This involves (a) filling the burette and allowing the liquid in it to drain until the meniscus is at a constant level, (b) adding a few drops of indicator solution to the titration flask, (c) reading the initial burette volume, (d) adding liquid to the titration flask from the burette a little at a time until the end-point is adjudged to have been reached, and (e) measuring the final level of liquid in the burette.

Even an elementary analysis of this type thus involves some ten separate steps, the last seven of which are normally, as we have seen, repeated several times. In principle, we should examine each step to evaluate the random and systematic errors that might occur. In practice, it is simpler to examine separately those stages which utilize weighings (steps 1(a) and 1(b)), and the remaining stages involving the use of volumetric equipment. (It is not intended to give detailed descriptions of the experimental techniques used in the various stages. Similarly, methods for calibrating weights, glassware, etc. will not be given.) Important amongst the contributions to the errors are the tolerances of the weights used in the gravimetric steps, and of the volumetric glassware. Standard specifications for these tolerances are issued by such bodies as the British Standards Institute (BSI) and the American Society for Testing and Materials (ASTM). The tolerance of a top-quality 100-g weight can be as low as ±0.25 mg, although for a weight used in routine work the tolerance would be up to four times as large. Similarly the tolerance for a grade A 250-ml standard flask is ±0.12 ml: grade B glassware generally has tolerances twice as large as grade A glassware. If a weight or a piece of glassware is within the tolerance limits, but not of exactly the correct weight or volume, a systematic error will arise. Thus, if the standard flask actually has a volume of 249.95 ml, this error will be reflected in the results of all the experiments based on the use of that flask. Repetition of the experiment will not reveal the error: in each replicate the volume will be assumed to be 250.00 ml when in fact it is less than this. If, however, the results of an experiment using this flask are compared with the results of several other experiments (e.g. in other laboratories) done with other flasks, then if all the flasks have slightly different volumes they will contribute to the random variation, i.e. the reproducibility, of the results.

Weighing procedures are normally associated with very small *random* errors. In routine laboratory work a 'four-place' balance is commonly used, and the random error involved should not be greater than ±0.0001–0.0002 g (the next chapter describes in detail the statistical terms used to express random errors). Given that the quantity being weighed is normally of the order of 1 g or more, it is evident that the random error, expressed as a percentage of the weight involved, is not more than 0.02%. A good standard material for volumetric analysis should (amongst other characteristics) have as high a formula weight as possible, in order to minimize these random weighing errors when a solution of a given molarity is being made up. In some analyses 'microbalances' are used to weigh quantities of a few milligrams – but the errors involved are likely to be only a few micrograms.

Systematic errors in weighings can be appreciable, and have a number of well established sources. These include adsorption of moisture on the surface of the weighing vessel; failure to allow heated vessels to cool to the same temperature as the balance before weighing (this error is especially commonplace in gravimetry when crucibles are weighed); corroded or dust-contaminated weights; and the buoyancy effect of the atmosphere, which acts to a different extent on objects of different density. For the most accurate work, weights must be calibrated against standards furnished by statutory bodies and standards authorities (see above). This calibration can be very accurate indeed, e.g. to ±0.01 mg for weights in the range 1–10 g. The buoyancy effect can be substantial. For example a sample of an organic liquid of density 0.92 g ml^{-1} which weighs 1.2100 g in air would weigh 1.2114 g *in vacuo*, a difference of more than 0.1%. Apart from the use of calibration procedures, which are further discussed in the next section, some simple experimental precautions can be taken to minimize these systematic errors. Weighing by difference (see above) cancels systematic errors arising from (for example) the moisture and other contaminants on the surface of the bottle (see also Section 2.12). If such precautions are taken, the errors in the weighing steps will be small, and it is probable that in most volumetric experiments weighing errors will be negligible compared with the errors arising from the use of volumetric equipment. Indeed, gravimetric methods are generally used for the calibration of an item of volumetric glassware, by weighing (in standard conditions) the water that it contains or delivers.

The *random* errors in volumetric procedures arise in the use of volumetric glassware. In filling a 250-ml standard flask to the mark, the error (i.e. the distance between the meniscus and the mark) might be about ±0.03 cm in a flask neck of diameter ca. 1.5 cm. This corresponds to a volume error of only about 0.05 ml – only 0.02% of the total volume of the flask. Similarly the random error in filling a 25-ml transfer pipette should not exceed 0.03 cm in a stem of diameter 0.5 cm; this gives a volume error of ca. 0.006 ml, 0.024% of the total volume. The error in reading a burette (of the conventional variety graduated in 0.1 ml divisions) is perhaps 0.01–0.02 ml. Each titration involves two such readings (the errors of which are *not* simply additive – see Chapter 2); if the titration volume is ca. 25 ml, the percentage error is again very small. The experimental conditions should be arranged so that the volume of titrant is not too small (say not less than 10 ml), otherwise the errors will become appreciable. (This precaution is analogous to choosing a standard compound of high formula weight to minimize the weighing error.) Even though a volumetric analysis involves several steps, in each of which a piece of volumetric glassware is used, it is apparent that the random errors should be small if the experiments are performed with care. In practice a good volumetric analysis should have a relative standard deviation (see Chapter 2) of not more than about 0.1%. Until recently such precision was not normally attainable in instrumental analysis methods, and it is still not common. In skilled hands, with all precautions taken, classical analysis methods can give results with relative standard deviations as low as 0.01%.

Volumetric procedures incorporate several important sources of systematic error. Chief amongst these are the drainage errors in the use of volumetric glassware, calibration errors in the glassware, and ‘indicator errors’. Perhaps the commonest error in routine volumetric analysis is to fail to allow enough time for a pipette to drain properly, or a meniscus level in a burette to stabilize. Moreover pipettes are of two types: those emptied by drainage, and blow-out pipettes from which the last remaining liquid must be forcibly expelled. To confuse the two types, for example by blowing out a drainage pipette, would certainly count as a gross error! Drainage errors have a systematic as well as a random effect: the volume delivered is invariably less than it should be. The temperature at which an experiment is performed has two effects. Volumetric equipment is conventionally calibrated at 20°C, but the temperature in an analytical laboratory may easily be several degrees different from this, and many experiments, for example in biochemical analysis, are carried out in 'cold rooms' at ca. 4°C. The temperature affects both the volume of the glassware and the density of liquids. The coefficient of expansion for dilute aqueous solutions is about 0.025% per degree, whereas a soda-glass vessel will change in volume by about 0.003% per degree and a borosilicate glass vessel by about 0.001% per degree. It is evident that the changes in the volumes of glassware will only be important in work of the highest quality, and even then only if the temperature is very different from 20°C. Furthermore, the effects of the expansion of the solutions will be largely self-cancelling if all the solutions are maintained at the same temperature. The effect is much more marked with non-aqueous solutions.

Indicator errors can be quite substantial – perhaps larger than the random errors in a typical titrimetric analysis. For example, in the titration of 0.1 M hydrochloric acid with 0.1 M sodium hydroxide, we expect the end-point to correspond to a pH of 7. In practice, however, we estimate this end-point by the use of an indicator such as methyl orange. Separate experiments show that this substance changes colour over the pH range ca. 3–4. If, therefore, the titration is performed by adding alkali to acid, the indicator will yield an apparent end-point when the pH is ca. 3.5, i.e. just before the true end-point. The systematic error involved here is likely to be as much as 0.2%. Conversely, if the titration is performed by adding acid to alkali, the end-point indicated by the methyl orange will actually be a little beyond the true end-point. In either case the error can be evaluated and corrected by performing a **blank** experiment, i.e. by determining how much alkali or acid is required to produce the indicator colour change in the *absence* of the acid (alkali).

In any analytical procedure, classical or instrumental, it should be possible to consider and estimate the sources of random and systematic error arising at each separate stage of the experiment. It is very desirable for the analyst to do this, in order to avoid major sources of error by careful experimental design (see Sections 1.5 and 1.6). It is worth noting, however, that titrimetric analyses are rather unusual in that they involve no single step having an error that is far greater than the errors in the other steps. In many analyses the overall error is in practice dominated by the error in a single step: this point is further discussed in the next chapter.

1.5 Handling systematic errors

Much of the remainder of this book will deal with the evaluation of random errors, which can be studied by a wide range of statistical methods. In most cases we shall assume for convenience that systematic errors are absent (though methods which test for the occurrence of systematic errors will be described). It is thus necessary at this stage to discuss systematic errors in more detail – how they arise, and how they may be countered. The example of the titrimetric analysis given in Section 1.3 shows clearly that systematic errors cause the mean value of a set of replicate measurements to deviate from the true value. It follows that (a) in contrast to random errors, systematic errors cannot be revealed merely by making repeated measurements, and that (b) unless the true result of the analysis is known in advance – an unlikely situation! – very large systematic errors might occur, but go entirely undetected unless suitable precautions are taken. In other words, it is all too easy to overlook substantial sources of systematic error. A small number of examples will clarify both the possible problems and their solutions.

In recent years, much interest has been shown in the levels of transition metals in biological samples such as blood serum. Many determinations have been made of the levels of (for example) chromium in serum – with startling results. Different workers, all studying pooled serum samples from healthy subjects, have obtained chromium concentrations varying from <1 to ca. 200 ng ml^{-1}. In general the lower results have been obtained more recently, and it has gradually become apparent that the earlier, higher values were due at least in part to contamination of the samples by chromium from stainless-steel syringes, tube caps, and so on. The determination of traces of chromium, for example by atomic-absorption spectrometry, is in principle relatively straightforward, and no doubt each group of workers achieved results which seemed satisfactory in terms of precision, but in a number of cases the large systematic error introduced by the contamination was entirely overlooked. Methodological systematic errors of this kind are extremely common – incomplete washing of a precipitate in gravimetric analysis, and the indicator error in volumetric analysis (see Section 1.4) are further well-known examples.

Another class of systematic error that occurs widely arises when false assumptions are made about the accuracy of an analytical instrument. Experienced analysts know only too well that the monochromators in spectrometers gradually go out of adjustment, so that errors of several nanometres in wavelength settings are not uncommon, yet many photometric analyses are undertaken without appropriate checks being made. Very simple devices such as volumetric glassware, stop-watches, pH-meters and thermometers can all show substantial systematic errors, but many laboratory workers regularly use these instruments as though they are always completely without bias. Moreover, the increasing availability of instruments controlled by microprocessors or microcomputers has reduced to a minimum the number of operations and the level of skill required of their operators. In these circumstances the temptation to regard the instruments' results as beyond reproach is overwhelming, yet such instruments (unless they are 'intelligent' enough to be self-calibrating – see Section 1.7) are still subject to systematic errors.

Systematic errors arise not only from procedures or apparatus; they can also arise from human bias. Some chemists suffer from astigmatism or colour-blindness (the latter is more common amongst men than women) which might introduce errors into their readings of instruments and other observations. A number of authors have reported various types of number bias, for example a tendency to favour even over odd numbers, or 0 and 5 over other digits, in the reporting of results. It is thus apparent that systematic errors of several kinds are a constant, and often hidden, risk for the analyst, so the most careful steps to minimize them must be considered.

Several different approaches to this problem are available, and any or all of them should be considered in each analytical procedure. The first precautions should be taken before any experimental work is begun. The analyst should consider carefully each stage of the experiment to be performed, the apparatus to be used and the sampling and analytical procedures to be adopted. At this earliest stage the likely sources of systematic error, such as the instrument functions that need calibrating, the steps of the analytical procedure where errors are most likely to occur, and the checks that can be made during the analysis, must be identified. Foresight of this kind can be immensely valuable (we shall see in the next section that similar advance attention should be given to the sources of random error) and is normally well worth the time invested. For example, a little thinking of this kind might well have revealed the possibility of contamination in the chromium determination described above.

The second line of defence against systematic errors lies in the design of the experiment at every stage. We have already seen (Section 1.4) that weighing by difference can remove some systematic gravimetric errors: these can be assumed to occur to the same extent in both weighings, so the subtraction process eliminates them. A further example of thoughtful experimental planning is provided by the spectrometer wavelength error described above. If the concentration of a sample of a single material is to be determined by absorption spectrometry, two procedures are possible. In the first, the sample is studied in a 1-cm path-length spectrometer cell at a single wavelength, say 400 nm, and the concentration of the test component is determined from the well-known equation $A = \varepsilon bc$ (where A, ε, c and b are respectively the measured absorbance, the accepted reference value of the molar absorptivity (units $\text{l mole}^{-1}\ \text{cm}^{-1}$) of the test component, the molar concentration of this analyte, and the path-length (cm) of the spectrometer cell). Several systematic errors can arise here. The wavelength might, as already discussed, be (say) 405 nm rather than 400 nm, thus rendering the reference value of ε inappropriate; this reference value might in any case be wrong; the absorbance scale of the spectrometer might exhibit a systematic error; and the path-length of the cell might not be exactly 1 cm. Alternatively, the analyst might take a series of solutions of the test substance of known concentration, and measure the absorbance of each at 400 nm. (One of these calibrating solutions would be a blank, i.e. it would contain all the analytical reagents but none of the test substance.) The results would then be used to construct a calibration graph for use in analysis of the test sample under exactly the same experimental conditions. This important approach to instrumental analysis is described in detail in Chapter 5. When this second method is used, the value of ε is not required, and the errors due to wavelength shifts, absorbance errors and path-length inaccuracies should cancel out, as they occur equally in the calibration and test experiments. Provided only that the conditions are indeed equivalent for the test and calibration samples (for example, that the same cell is used and the wavelength and absorbance scales do not alter *during the experiment*) all the major sources of systematic error are in principle eliminated.

The final and perhaps most formidable protection against systematic errors is the use of standard reference materials and methods. Before the experiment is started, each piece of apparatus is calibrated by an appropriate procedure. We have seen that volumetric equipment can be calibrated by the use of gravimetric methods. Similarly, spectrometer wavelength scales can be calibrated with the aid of standard light-sources which have narrow emission lines at well-established wavelengths, and spectrometer absorbance scales can be calibrated with standard solid or liquid filters. In analogous fashion, most pieces of equipment can be calibrated so that their systematic errors are known in advance. The importance of this area of chemistry (and other experimental sciences) is reflected in the extensive work of bodies such as the National Physical Laboratory and LGC (Laboratory of the Government Chemist) (in the UK), the National Institute for Science and Technology (NIST) (in the USA) and similar organizations elsewhere. Whole volumes have been written on the standardization of particular types of equipment, and a number of commercial organizations specialize in the sale of standard reference materials.

If systematic errors are occurring during chemical processes or as a result of using impure reagents, rather than in the equipment, an alternative form of comparison must be used, i.e. the determination must be repeated by use of an entirely independent procedure. If two (or more) chemically and physically unrelated methods are used to perform one analysis, and if they consistently yield results showing only random differences, it is a reasonable presumption that no significant systematic errors are present. For this approach to be valid, *each step* of the two experiments has to be independent. Thus in the case of serum chromium determinations, it would not be sufficient to replace the atomic-absorption spectrometry step by a colorimetric method or by plasma spectrometry. The systematic errors would only be revealed by altering the sampling methods also, e.g. by minimizing or eliminating the use of stainless-steel equipment. A further important point is that comparisons must be made over the whole of the concentration range for which an analytical procedure is to be used. For example, the bromocresol green dye-binding method for the determination of albumin in serum correlates well with alternative methods (e.g. immunological methods) at normal or high levels of albumin, but when the albumin levels are abnormally low (these are inevitably the cases of most clinical interest!) the agreement between the two methods is poor, the dye-binding method giving consistently (and erroneously) higher albumin concentrations. The statistical approaches used in method comparisons are described in detail in Chapters 3 and 5.

The prevalence of systematic errors in everyday analytical work is well illustrated by the results of **collaborative trials**. If an able and experienced analyst finds 10 ng ml^{-1} of a drug in a urine sample, it is natural to suppose that other analysts would obtain closely similar results for the same sample, any differences being due to random errors only. Unfortunately, this is far from true in practice. Many collaborative studies involving different laboratories, when aliquots of a single sample are examined by the same experimental procedures and types of instrument, show variations in the results much greater than those expected from random errors. The inescapable conclusion is that in many laboratories substantial systematic errors, both positive and negative, are going undetected or uncorrected. The obvious importance of this situation, which has serious implications for all analytical scientists, has encouraged many studies of the methodology of collaborative trials and **proficiency schemes**, and of the statistical evaluation of their results. Such schemes have recently led to dramatic improvements in the quality of analytical results in many fields. These topics are discussed in more detail in Chapter 4.

1.6 Planning and design of experiments

Many chemists regard statistical tests as methods to be used only to assess the results of completed experiments. While this is indeed a crucial area of application of statistics, we must also be aware of the importance of statistical concepts in the planning and design of experiments. In the previous section the value of trying to predict systematic errors in advance, thereby permitting the analyst to lay plans for countering them, was emphasized. The same considerations apply to random errors. As will be seen in Chapter 2, the combination of the random errors of the individual parts of an experiment to give an overall error requires the use of simple statistical formulae. In practice, the overall error is often dominated by the error in just one stage of the experiment, other errors having negligible effects when all the errors are combined correctly. Again it is obviously desirable to try to identify, *before the experiment begins*, where this single dominant error is likely to arise, and then to try to minimize it. Although random errors can never be eliminated, they can certainly be minimized by particular attention to experimental techniques: improving the precision of a spectrometric experiment by using a constant temperature sample cell would be a simple instance of such a precaution. For both random and systematic errors, therefore, the moral is clear: every effort must be made to identify the serious sources of error before practical work starts, so that the experiment can be designed to minimize such errors.

There is another and more subtle aspect of experimental design. In many analyses, one or more of the desirable features of the method (for example sensitivity, selectivity, sampling rate, low cost, etc.) will be found to depend on a number of experimental factors. We shall wish to design the experiment so that the optimum combination of these factors is used, thereby obtaining the best sensitivity, selectivity, etc. Although some preliminary experiments and prior knowledge will be involved, optimization should be performed (in the interests of conserving resources and time) before the method is put into routine or widespread use.

The complexity of optimization procedures can be illustrated with the aid of an example. In enzymatic analyses, the concentration of the analyte is determined from observations of the rate of an enzyme-catalysed reaction. The analyte is often the substrate, i.e. the compound that is changed in the reaction catalysed by the enzyme. Let us assume that we are seeking the maximum reaction rate in a particular experiment, and that the rate in practice depends upon (amongst other factors) the pH of the reaction mixture, the temperature, and the concentration of the enzyme. How are the optimum conditions to be found? It is easy to identify one possible approach. The analyst could perform a series of experiments, in each of which the enzyme concentration and the temperature are kept constant but the pH is varied. In each case the rate of the enzyme-catalysed reaction would be determined and an optimum pH value would thus be obtained – let us say that it is 7.5. A second series of reaction-rate experiments could then be performed, with the pH maintained at 7.5, the enzyme concentration again fixed, but the temperature varied. An optimum temperature would thus be found, say 40°C. Finally a series of experiments at pH 7.5 and 40°C but with various enzyme concentrations would indicate the optimum enzyme level. This approach to the optimization of the experiment is clearly tedious: in more realistic examples many more than three experimental factors might need investigation. There is the further and fundamental objection that the method assumes that the factors (pH, temperature, enzyme concentration) affect the reaction rate in an *independent* way. This might not be true. For example we have found that, at pH 7.5, the optimum temperature is 40°C. At a different pH, however, the optimum temperature might not be 40°C. In other words these factors may affect the reaction rate in an *interactive* way, and the conditions established in the series of experiments just described might not actually be the optimum ones. Had the first series of experiments been started in different conditions, a different set of 'optimum' values might have been obtained. It is apparent from this simple example that experimental optimization can be a formidable problem. This very important aspect of statistics as applied to analytical chemistry is considered in more detail in Chapter 7.

1.7 Calculators and computers in statistical calculations

No chemist can be unaware of the astonishing developments that have recently taken place in the realm of microelectronics. These advances have made possible the construction of devices that enormously simplify statistical calculations. The rapid growth of **chemometrics** – the application of mathematical methods to the solution of chemical problems of all types – is due to the ease with which large quantities of data can be handled, and complex calculations done, with calculators and computers.

These devices are available to the analytical chemist at several levels of complexity and cost. Hand-held calculators are extremely cheap, very reliable, and capable of performing many of the routine statistical calculations described in this book with a minimal number of keystrokes. Pre-programmed functions allow calculations of mean and standard deviation (see Chapter 2) and correlation and linear regression (see Chapter 5). Other calculators can be programmed by the user to perform additional calculations such as confidence limits (see Chapter 2), significance tests (see Chapter 3) and non-linear regression (see Chapter 5). For many applications in laboratories performing analytical research or routine analyses, calculators of these types will be more than adequate. Their only disadvantage is their inability to handle very large quantities of data.

Personal computers (PCs) are now found in all chemical laboratories. Most modern instruments are controlled by PCs, which also handle and report the analytical data obtained, and most analytical scientists have their own PCs on their desks. Portable PCs facilitate the recording and calculation of data in the field, and are readily linked to their larger cousins on returning to the laboratory. Many instruments are now devoid of manual controls, because they are entirely controlled by a linked PC. Additional functions of the PC can include checking instrument performance, diagnosing and reporting malfunctions, storing large databases (e.g. of digitized spectra) and comparing analytical data with the databases, optimizing operating conditions (see Chapter 7), and selecting and using a variety of calibration calculations. One concern is that the software provided with computer-controlled instruments is not always explained to the user: an analyst might thus suffer the misfortune of having a data set interpreted via (for example) a calibration routine which was not defined and might not always be appropriate. This is an undesirable situation, though the desire of instrument companies to try to protect costly dedicated software from piracy is understandable.

A wealth of excellent general statistical software is available for PCs. The memory capacity and speed of the computers are now more than adequate for work with all but the largest data sets, and PCs are routinely provided with word processors, which greatly aid the compilation of analytical reports and papers. Also universally available are **spreadsheet** programs. These, though originally designed for financial calculations, are often more than adequate for statistical work, having many built-in statistical functions and excellent graphical presentation facilities. The popularity of spreadsheets derives from their speed and simplicity in use, and their ability to perform almost instant 'what if' calculations: for example, what would the mean and standard deviation of a set of results be if one suspect piece of data is omitted? Spreadsheets are designed to facilitate rapid data entry, and data in spreadsheet format can easily be exported to the more specialist suites of statistics software. **Microsoft Excel** is the most widely used spreadsheet, and offers most of the statistical facilities that many users of this book should need. A number of examples of its application are provided in later chapters, and the bibliography lists some books which describe and enhance its application to statistical problems.

More advanced calculation facilities are provided by specialized suites of statistical software. Amongst these **Minitab** is very widely used in educational establishments and research laboratories. In addition to the expected simple statistical functions it offers many more advanced calculations, including multivariate methods (see Chapter 8), initial data analysis (EDA) and non-parametric tests (see Chapter 6), experimental design (see Chapter 7) and many quality control methods (Chapter 4). More specialized and excellent programs for various types of multivariate analysis are also available for PCs: the best known is The Unscrambler. New and updated versions of these programs, with extra facilities and/or improved user interfaces, appear at regular intervals. Although some help facilities are available in every case, it is fair to say that the software is usually designed for users rather than students, and does not have a strongly tutorial emphasis. But a program specifically designed for tutorial purposes, VAMSTAT, is a valuable tool, with on-screen tests for students and clear explanations of many important methods.

Yet another advantageous feature of PCs is that they can readily be 'networked', i.e. a group of PCs in the same or neighbouring laboratories can be linked so that both operating software and data can be freely passed from one to another. An obvious benefit of such networks is in the establishment of Laboratory Information Management Systems (LIMS) which allow large numbers of analytical specimens to be identified and tracked as they move through a laboratory. Samples are identified and tracked by barcoding or similar systems, and the PCs attached to a range of instruments send a range of analytical results to a central computer which (for example) prints a summary report, including a statistical evaluation.

It is most important for the analytical chemist to remember that the availability of all these data-handling facilities increases rather than decreases the need for a sound knowledge of the principles underlying statistical calculations. A computer or calculator will rapidly perform any statistical test or calculation selected by the user, *whether or not that procedure is suitable for the data under study*. For example, a linear least squares program will determine a straight line to fit *any* set of x and y values, even in cases where visual inspection would show at once that such a program is entirely inappropriate (see Chapter 5). Again, a simple program for testing the significance of the difference between the means of two data sets may assume that the variances (see Chapter 2) of the two sets are similar: but the program will blindly perform the calculation on request and provide a 'result' even if the variances actually differ significantly. Even quite comprehensive suites of computer programs often fail to provide advice on the choice of statistical method appropriate to a given set of data. The analyst must thus use both a knowledge of statistics and common sense to ensure that the correct calculation is performed.

Bibliography

[NB – The following books are likely to be of value as support material for most of the chapters in this book.]

Anderson, R. L. 1987. *Practical Statistics for Analytical Chemistry*. Van Nostrand Reinhold, New York.

British Standards Institute. 1987. *Schedule for Tables for Use in the Calibration of Volumetric Glassware*. [BS1797]. BSI, London. (Valuable example of a BSI monograph.)

Diamond, D. and Hanratty, V. C. 1997. *Spreadsheet Applications in Chemistry Using Microsoft Excel*. Wiley, New York. (Clear guidance on the use of Excel, with many examples from analytical and physical chemistry.)

Meier, P. C. and Zund, R. E. 1993. *Statistical Methods in Analytical Chemistry*. Wiley, New York. (An advanced text, with software and data sets provided on disk.)

Middelton, M. R. 1997. *Data Analysis Using Microsoft Excel*, Duxbury, Belmont. (A clear guide to the very wide range of statistical procedures available in Excel.)

Minitab Reference Manual, Version 11. June 1996. Minitab Inc, State College, USA.

Neave, H. R. 1981. *Elementary Statistics Tables*. Routledge, London. (All users of statistics require a good set of statistical tables: this set is strongly recommended, in view of its clear explanations of the tables and the associated statistical tests.)

Youden, W. J. and Steiner, E. H. 1975. *Statistical Manual of the Association of Official Analytical Chemists*. AOAC, Arlington. (Classic manual on precision and accuracy, and on collaborative studies.)

Exercises

1. A standard sample of pooled human blood serum contains 42.0 g of albumin per litre. Five laboratories (A–E) each do six determinations (on the same day) of the albumin concentration, with the following results ($g\,l^{-1}$ throughout):

A	42.5	41.6	42.1	41.9	41.1	42.2
B	39.8	43.6	42.1	40.1	43.9	41.9
C	43.5	42.8	43.8	43.1	42.7	43.3
D	35.0	43.0	37.1	40.5	36.8	42.2
E	42.2	41.6	42.0	41.8	42.6	39.0

   Comment on the bias, precision and accuracy of each of these sets of results.

2. Using the same sample and method as in question 1, laboratory A makes six further determinations of the albumin concentration, this time on six successive days. The values obtained are 41.5, 40.8, 43.3, 41.9, 42.2 and 41.7 $g\,l^{-1}$. Comment on these results.

3. The number of binding sites per molecule in a sample of monoclonal antibody is determined four times, with results of 1.95, 1.95, 1.92 and 1.97. Comment on the bias, precision and accuracy of these results.

4. Discuss the degrees of bias and precision desirable or acceptable in the following analyses:

   (i) Determination of the lactate concentration of human blood samples.
   (ii) Determination of uranium in an ore sample.
   (iii) Determination of a drug in blood plasma after an overdose.
   (iv) Study of the stability of a colorimetric reagent by determination of its absorbance at a single wavelength over a period of several weeks.

5. For each of the following experiments, try to identify the major probable sources of random and systematic errors, and consider how such errors may be minimized:

   (i) The iron content of a large lump of ore is determined by taking a single small sample, dissolving it in acid, and titrating with ceric sulphate after reduction of Fe(III) to Fe(II).
   (ii) The same sampling and dissolution procedure is used as in (i) but the iron is determined colorimetrically after addition of a chelating reagent and extraction of the resulting coloured complex into an organic solvent.
   (iii) The sulphate content of an aqueous solution is determined gravimetrically with barium chloride as the precipitant.

CHAPTER TWO

Statistics of repeated measurements

2.1 Mean and standard deviation

In Chapter 1 we saw that it is usually necessary to make repeated measurements in many analytical experiments in order to reveal the presence of random errors. This chapter applies some fundamental statistical concepts to such a situation. We will start by looking again at the example in Chapter 1 which considered the results of five replicate titrations done by each of four students. These results are reproduced below.

Student			*Results (ml)*		
A	10.08	10.11	10.09	10.10	10.12
B	9.88	10.14	10.02	9.80	10.21
C	10.19	9.79	9.69	10.05	9.78
D	10.04	9.98	10.02	9.97	10.04

Two criteria were used to compare these results, the average value (technically known as a measure of location) and the degree of spread (or dispersion). The average value used was the **arithmetic mean** (usually abbreviated to the **mean**), which is the sum of all the measurements divided by the number of measurements.

The mean, $\bar{x}$, of n measurements is given by $\bar{x} = \frac{\sum x_i}{n}$ (2.1)

In Chapter 1 the spread was measured by the difference between the highest and lowest values (the **range**). A more useful measure, which utilizes all the values, is the **standard deviation**, s, which is defined as follows:

The standard deviation, s, of n measurements is given by

$$s = \sqrt{\sum_i (x_i - \bar{x})^2/(n-1)} \tag{2.2}$$

The calculation of these statistics can be illustrated by an example.

EXAMPLE 2.1.1

Find the mean and standard deviation of A's results.

	x_i	$(x_i - \bar{x})$	$(x_i - \bar{x})^2$
	10.08	−0.02	0.0004
	10.11	0.01	0.0001
	10.09	−0.01	0.0001
	10.10	0.00	0.0000
	10.12	0.02	0.0004
Totals	50.50	0	0.0010

$$\bar{x} = \frac{\sum x_i}{n} = \frac{50.50}{5} = 10.1 \text{ ml}$$

$$s = \sqrt{\sum_i (x_i - \bar{x})^2/(n-1)} = \sqrt{0.001/4} = 0.0158 \text{ ml}$$

Note that $\sum(x_i - \bar{x})$ is always equal to 0.

The answers to this example have been arbitrarily given to three significant figures: further discussion of this important aspect of the presentation of results is considered in Section 2.8. The reader can check that the standard deviations of the results of students B, C and D are 0.172, 0.210, and 0.0332 ml respectively, giving quantitative confirmation of the assessments of precision made in Chapter 1.

In practice, it is most unusual to make these calculations on paper. All except the most basic pocket calculators will give the results if the values of x_i are keyed in. However, care must be taken that the correct key is pressed to obtain the standard deviation. Some calculators give two different values for the standard deviation, one calculated by using equation (2.2) and the other calculated with the denominator of this equation, i.e. $(n - 1)$, replaced by n. (The reason for these two different forms is explained on p. 23.) Obviously, for large values of n the difference is negligible. Alternatively, readily available computer software can be used to perform these calculations (see Chapter 1).

The square of s is a very important statistical quantity known as the **variance**; its value will become apparent later in this chapter when we consider the propagation of errors.

Variance = the square of the standard deviation, s^2

Another widely used measure of spread is the **coefficient of variation (CV)**, also known as the **relative standard deviation (RSD)**, which is given by $100s/\bar{x}$.

Coefficient of variation (CV) = relative standard deviation (RSD) = $100\,s/\bar{x}$.

The CV or RSD, the units of which are obviously per cent, is an example of a **relative error**, i.e. an error estimate divided by an estimate of the absolute value of the measured quantity. Relative errors are frequently used in the comparison of the precision of results which have different units or magnitudes, and are again important in calculations of error propagation.

2.2 The distribution of repeated measurements

Although the standard deviation gives a measure of the spread of a set of results about the mean value, it does not indicate the shape of the distribution. To illustrate this we need a large number of measurements such as those in Table 2.1. This gives the results of 50 replicate determinations of the nitrate ion concentration in a particular water specimen, given to two significant figures.

These results can be summarized in a **frequency table** (Table 2.2). This table shows that, in Table 2.1, the value 0.46 $\mu g\,ml^{-1}$ appears once, the value 0.47 $\mu g\,ml^{-1}$ appears three times and so on. The reader can check that the mean of these results is 0.500 $\mu g\,ml^{-1}$ and the standard deviation is 0.0165 $\mu g\,ml^{-1}$. The distribution of the results can most easily be appreciated by drawing a **histogram** as in Figure 2.1. This shows that the distribution of the measurements is roughly symmetrical about the mean, with the measurements clustered towards the centre.

This set of 50 measurements is a sample from the very large (in theory infinite) number of measurements which we could make of the nitrate ion concentration. The set of all possible measurements is called the **population**. *If there are no systematic errors*, then the mean of this population, denoted by

Table 2.1 Results of 50 determinations of nitrate ion concentration, in $\mu g\,ml^{-1}$

0.51	0.51	0.51	0.50	0.51	0.49	0.52	0.53	0.50	0.47
0.51	0.52	0.53	0.48	0.49	0.50	0.52	0.49	0.49	0.50
0.49	0.48	0.46	0.49	0.49	0.48	0.49	0.49	0.51	0.47
0.51	0.51	0.51	0.48	0.50	0.47	0.50	0.51	0.49	0.48
0.51	0.50	0.50	0.53	0.52	0.52	0.50	0.50	0.51	0.51

Table 2.2 Frequency table for measurements of nitrate ion concentration

Nitrate ion concentration ($\mu g\,ml^{-1}$)	*Frequency*
0.46	1
0.47	3
0.48	5
0.49	10
0.50	10
0.51	13
0.52	5
0.53	3

μ, is the true value of the nitrate ion concentration which we are trying to determine. The mean of the sample gives us an estimate of μ. Similarly, the population has a standard deviation, denoted by σ. The value of the standard deviation, s, of the sample gives us an estimate of σ. Use of equation (2.2) gives us an unbiased estimate of σ. If n, rather than $(n - 1)$, is used in the denominator of the equation the value of s obtained tends to underestimate σ (see p. 21).

The measurements of nitrate ion concentration given in Table 2.2 have only certain discrete values, because of the limitations of the method of measurement. In theory a concentration could take any value, so a continuous curve is needed to describe the form of the population from which the sample was taken. The mathematical model usually used is the **normal** or **Gaussian distribution** which is described by the equation

$$y = \frac{1}{\sigma\sqrt{2\pi}} \exp\{-(x - \mu)^2/2\sigma^2\} \qquad (2.3)$$

Its shape is shown in Figure 2.2. There is no need to remember this complicated formula, but some of its general properties are important. The

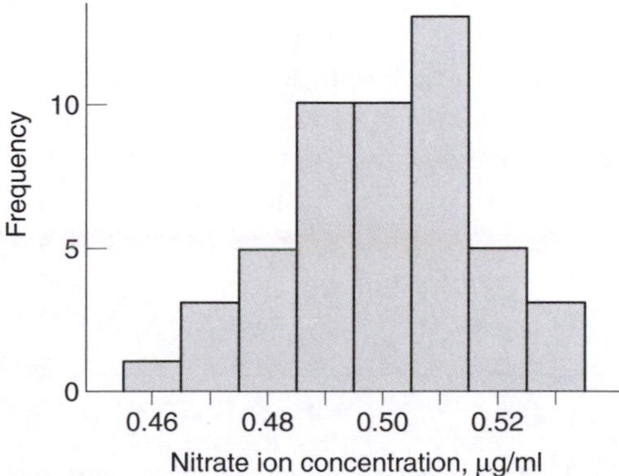

Figure 2.1 Histogram of the nitrate ion concentration data in Table 2.2.

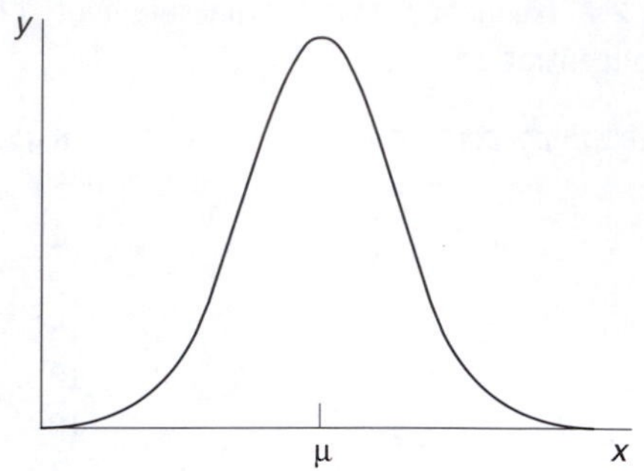

Figure 2.2 The normal distribution, $y = \exp[-(x - \mu)^2/2\sigma^2]/\sigma\sqrt{2\pi}$. The mean is indicated by μ.

curve is symmetrical about μ and the greater the value of σ the greater the spread of the curve, as shown in Figure 2.3. More detailed analysis shows that, whatever the values of μ and σ, the normal distribution has the following properties.

> For a normal distribution with mean μ and standard deviation σ, approximately 68% of the population values lie within $\pm 1\sigma$ of the mean, approximately 95% of the population values lie within $\pm 2\sigma$ of the mean, and approximately 99.7% of the population values lie within $\pm 3\sigma$ of the mean.

These properties are illustrated in Figure 2.4. This would mean that, if the nitrate ion concentrations (in $\mu g\, ml^{-1}$) given in Table 2.2 are normally distributed, then about 68% should lie in the range 0.483–0.517, about 95% in the range 0.467–0.533 and 99.7% in the range 0.450–0.550. In fact 33 of the 50

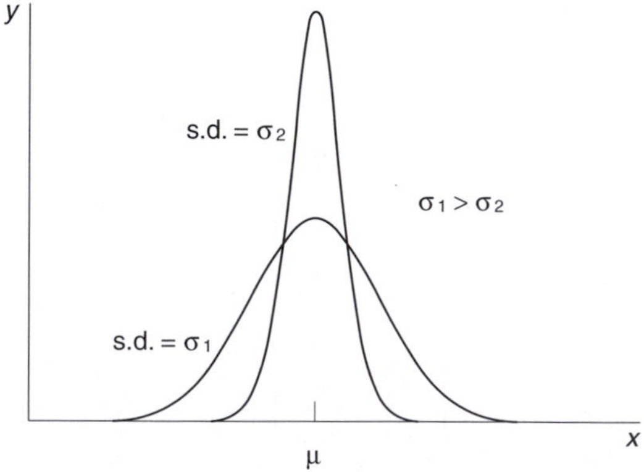

Figure 2.3 Normal distributions with the same mean but different values of the standard deviation.

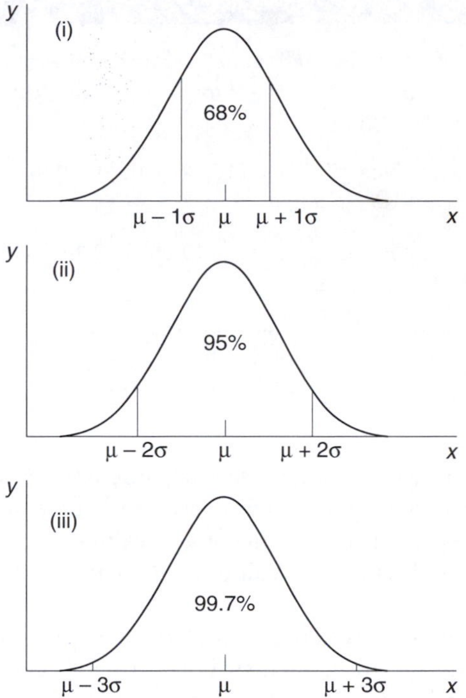

Figure 2.4 Properties of the normal distribution: (i) approximately 68% of values lie within $\pm 1\sigma$ of the mean; (ii) approximately 95% of values lie within $\pm 2\sigma$ of the mean; (iii) approximately 99.7% of values lie within $\pm 3\sigma$ of the mean.

results (66%) lie between 0.483 and 0.517, 49 (98%) between 0.467 and 0.533, and all the results between 0.450 and 0.550, so the agreement with theory is fairly good.

For a normal distribution with known mean, μ, and standard deviation, σ, the exact proportion of values which lie within any interval can be found from tables, provided that the values are first **standardized** so as to give ***z*-values**. This is done by expressing a value of x in terms of its deviation from the mean in units of standard deviation, σ. That is

$$\text{Standardized normal variable, } z = \frac{(x - \mu)}{\sigma} \tag{2.4}$$

Table A.1 (Appendix 2) gives the proportion of values, $F(z)$, that lie below a given value of z. $F(z)$ is called the **standard normal cumulative distribution function**. For example the proportion of values below $z = 2$ is $F(2) = 0.9772$ and the proportion of values below $z = -2$ is $F(-2) = 0.0228$. Thus the *exact* value for the proportion of measurements lying within 2 standard deviations of the mean is $0.9772 - 0.0228 = 0.9544$.

EXAMPLE 2.2.1

If repeated values of a titration are normally distributed with mean 10.15 ml and standard deviation 0.02 ml, find the proportion of measurements which lie between 10.12 ml and 10.20 ml.

Standardizing the first value gives $z = (10.12 - 10.15)/0.02 = -1.5$.
From Table A.1, $F(-1.5) = 0.0668$.

Standardizing the second value gives $z = (10.20 - 10.15)/0.02 = 2.5$.
From Table A.1, $F(2.5) = 0.9938$.

Thus the proportion of values between $x = 10.12$ to 10.20 (which corresponds to $z = -1.5$ to 2.5) is $0.9938 - 0.0668 = 0.927$.

The reader should be warned that there is considerable variation in the format of tables for calculating proportions from z values. Some tables only give positive z values, so the proportions for negative z values then have to be deduced using considerations of symmetry. Values of $F(z)$ can also be found using Excel or Minitab.

Although it cannot be proved that repeated measurements of a single analytical quantity are always normally distributed, there is considerable evidence that this assumption is generally at least approximately true. Also, we shall see when we come to look at sample means that any departure of a population from normality is not usually important in the context of the statistical tests most frequently used.

The normal distribution is not only applicable to repeated measurements made on the same specimen. It also often fits the distribution of results obtained when the same quantity is measured for different materials from similar sources. For example, if we measured the concentration of albumin in blood sera taken from healthy adult humans we would find the results were approximately normally distributed.

2.3 Log-normal distribution

In situations where one measurement is made on each of a number of specimens, distributions other than the normal distribution can also occur. In particular the so-called **log-normal distribution** is frequently encountered. For this distribution, frequency plotted against the *logarithm* of the concentration (or other characteristics) gives a normal distribution curve. An example of a variable which has a log-normal distribution is the antibody concentration in human blood sera. When frequency is plotted against concentration for this variable, the asymmetrical curve shown in Figure 2.5a is obtained. If, however, the frequency is plotted against the logarithm (e.g. to the base 10) of the concentration, an approximately normal distribution is obtained, as shown in Figure 2.5b. Another

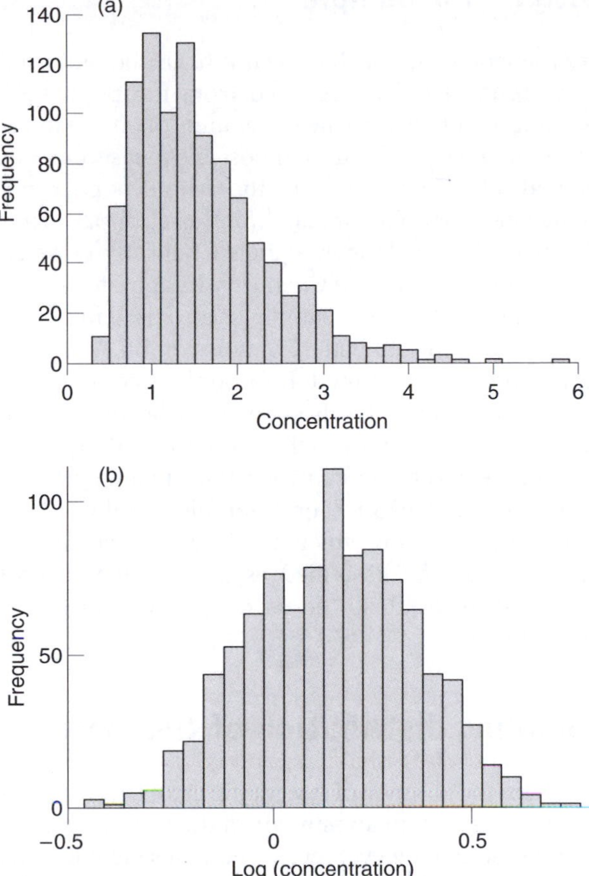

Figure 2.5 (a) An approximately log-normal distribution: concentration of serum immunoglobulin M antibody in male subjects. (b) The results in (a) plotted against the logarithm of the concentration.

example of a variable which may follow a log-normal distribution is the particle size of the droplets formed by the nebulizers used in flame spectroscopy.

The interval containing a given percentage of the measurements for a variable which is log-normally distributed can be found by working with the logarithms of the values. The distribution of the logarithms of the blood serum concentration shown in Figure 2.5b has mean 0.15 and standard deviation 0.20. This means that approximately 68% of the logged values lie in the interval $0.15 - 0.20$ to $0.15 + 0.20$, that is -0.05 to 0.35. Taking antilogarithms we find that 68% of the original measurements lie in the interval $10^{-0.05}$ to $10^{0.35}$, that is 0.89 to 2.24. The antilogarithm of the mean of the logged values, $10^{0.15} = 1.41$, gives the *geometric* mean of the original distribution where the geometric mean is given by $\sqrt[n]{x_1 x_2 \ldots x_n}$.

2.4 Definition of a 'sample'

In this chapter the word 'sample' has been introduced and used in its statistical sense of a group of objects selected from the population of all such objects, for example a sample of 50 measurements of nitrate ion concentration from the (infinite) population of all such possible measurements, or a sample of healthy human adults chosen from the whole population in order to measure the concentration of serum albumin for each one. The Commission on Analytical Nomenclature of the Analytical Chemistry Division of the International Union of Pure and Applied Chemistry has pointed out that confusion and ambiguity can arise if the term 'sample' is also used in its colloquial sense of the 'actual material being studied' (Commission on Analytical Nomenclature, 1990). It recommends that the term sample is confined to its statistical concept. Other words should be used to describe the material on which measurements are being made, in each case preceded by 'test', for example **test solution** or **test extract**. We can then talk unambiguously of a sample of measurements on a test extract, or a sample of tablets from a batch. A test portion from a population which varies with time, such as a river or circulating blood, should be described as a **specimen**. Unfortunately this practice is by no means usual, so the term 'sample' remains in use for two related but distinct uses.

2.5 The sampling distribution of the mean

We have seen that, in the absence of systematic errors, the mean of a sample of measurements provides us with an estimate of the true value, μ, of the quantity we are trying to measure. However, even in the absence of systematic errors, the individual measurements vary due to random errors and so it is most unlikely that the mean of the sample will be *exactly* equal to the true value. For this reason it is more useful to give a range of values which is likely to include the true value. The width of this range depends on two factors. The first is the precision of the individual measurements, which in turn depends on the standard deviation of the population. The second is the number of measurements in the sample. The very fact that we repeat measurements implies we have more confidence in the mean of several values than in a single one. Intuitively we would expect that the more measurements we make, the more reliable our estimate of μ, the true value, will be.

To pursue this idea, let us return to the nitrate ion determination described in Section 2.2. In practice it would be most unusual to make 50 repeated measurements in such a case: a more likely number would be five. We can see how the means of samples of this size are spread about μ by treating the results in Table 2.2 as ten samples, each containing five results. Taking each column as one sample, the means are 0.506, 0.504, 0.502, 0.496, 0.502, 0.492, 0.506, 0.504, 0.500, 0.486. We can see that these means are more closely clustered than the original measurements. If we continued to take samples of five measurements and calculated their means, these means would have a frequency distribution of their own. The distribution of all possible sample means (in this case an infinite number) is called the **sampling distribution of the mean**. Its mean is the same as the mean of the original population. Its standard deviation is called the **standard error of the mean** (s.e.m.). There is an exact mathematical relationship between the latter and the standard deviation, σ, of the distribution of the individual measurements:

> For a sample of n measurements,
>
> $$\text{standard error of the mean (s.e.m.)} = \sigma/\sqrt{n} \qquad (2.5)$$

As expected, the larger n is, the smaller the value of the s.e.m. and consequently the smaller the spread of the sample means about μ.

The term, 'standard error of the mean', might give the impression that $\sigma/\sqrt{n}$ gives the difference between μ and $\bar{x}$. This is not so: $\sigma/\sqrt{n}$ gives a measure of the variability of $\bar{x}$, as we shall see in the next section.

Another property of the sampling distribution of the mean is that, *even if the original population is not normal*, the sampling distribution of the mean tends to the normal distribution as n increases. This result is known as the **central limit theorem**. This theorem is of great importance because many statistical tests are performed on the mean and assume that it is normally distributed. Since in practice we can assume that distributions of repeated measurements are at least approximately normally distributed, it is reasonable to assume that the means of quite small samples (say >5) are normally distributed.

2.6 Confidence limits of the mean for large samples

Now that we know the form of the sampling distribution of the mean we can return to the problem of using a sample to define a range which we may reasonably assume includes the true value. (Remember that in doing this we are assuming systematic errors to be absent.) Such a range is known as a **confidence interval** and the extreme values of the range are called the **confidence limits**. The term 'confidence' implies that we can assert with a given degree of confidence, i.e. a certain probability, that the confidence interval does include the true value. The size of the confidence interval will obviously depend on how certain we want to be that it includes the true value: the greater the certainty, the greater the interval required.

Figure 2.6 shows the sampling distribution of the mean for samples of size n. If we assume that this distribution is normal then 95% of the sample means will lie in the range given by:

$$\mu - 1.96(\sigma/\sqrt{n}) < \bar{x} < \mu + 1.96(\sigma/\sqrt{n}) \qquad (2.6)$$

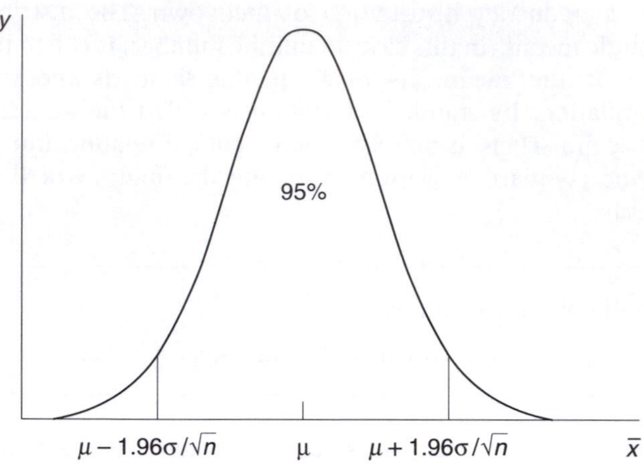

Figure 2.6 The sampling distribution of the mean, showing the range within which 95% of sample means lie.

(The exact value 1.96 has been used in this equation rather than the approximate value, 2, quoted in Section 2.2. The reader can use Table A.1 to check that the proportion of values between $z = -1.96$ and $z = 1.96$ is indeed 0.95.)

In practice, however, we usually have one sample, of known mean, and we require a range for μ, the true value. Equation (2.6) can be rearranged to give this:

$$\bar{x} - 1.96(\sigma/\sqrt{n}) < \mu < \bar{x} + 1.96(\sigma/\sqrt{n}) \tag{2.7}$$

Equation (2.7) gives the **95% confidence interval of the mean**. The **95% confidence limits** are $\bar{x} \pm 1.96\sigma/\sqrt{n}$.

In practice we are unlikely to know σ exactly. However, *provided that the sample is large*, σ can be replaced by its estimate, s.

Other confidence limits are sometimes used, in particular the 99% and 99.7% confidence limits.

For large samples, the confidence limits of the mean are given by

$$\bar{x} \pm zs/\sqrt{n} \tag{2.8}$$

where the value of z depends on the degree of confidence required.

For 95% confidence limits, $z = 1.96$

For 99% confidence limits, $z = 2.58$

For 99.7% confidence limits, $z = 2.97$

EXAMPLE 2.6.1

Calculate the 95% and 99% confidence limits of the mean for the nitrate ion concentration measurements in Table 2.1.

We have $\bar{x} = 0.500$, $s = 0.0165$ and $n = 50$. Using equation (2.8) gives the 95% confidence limits as:

$$\bar{x} \pm 1.96s/\sqrt{n} = 0.500 \pm 1.96 \times 0.0165/\sqrt{50} = 0.500 \pm 0.0046\,\mu\text{g ml}^{-1}$$

and the 99% confidence limits as:

$$\bar{x} \pm 2.58s/\sqrt{n} = 0.500 \pm 2.58 \times 0.01651/\sqrt{50} = 0.500 \pm 0.0060\,\mu\text{g ml}^{-1}$$

In this example it is interesting to note that although the original measurements varied between 0.46 and 0.53, the 99% confidence interval for the mean is from 0.494 to 0.506.

2.7 Confidence limits of the mean for small samples

As the sample size gets smaller, s becomes less reliable as an estimate of σ. This can be seen by again treating each column of the results in Table 2.2 as a sample of size five. The standard deviations of the ten columns are 0.009, 0.015, 0.026, 0.021, 0.013, 0.019, 0.013, 0.017, 0.010, 0.018. We see that the largest value of s is nearly three times the size of the smallest. To allow for this, equation (2.8) must be modified.

For small samples, the confidence limits of the mean are given by

$$\bar{x} \pm t_{n-1}s/\sqrt{n} \tag{2.9}$$

The subscript $(n - 1)$ indicates that t depends on this quantity, which is known as the number of **degrees of freedom, d.f.** (usually given the symbol ν). [The term 'degrees of freedom' refers to the number of *independent* deviations $(x_i - \bar{x})$ which are used in calculating s. In this case the number is $(n - 1)$, because when $(n - 1)$ deviations are known the last can be deduced since $\sum_i (x_i - \bar{x}) = 0$]. The value of t also depends on the degree of confidence required. Some values of t are given in Table 2.3. A more complete version of this table is given in Table A.2 in Appendix 2.

For large n, the values of t_{n-1} for confidence intervals of 95% and 99% respectively are very close to the values 1.96 and 2.58 used in Example 2.6.1. The following example illustrates the use of equation (2.9).

Table 2.3 Values of t for confidence intervals

Degrees of freedom	*Values of t for confidence interval of*	
	95%	*99%*
2	4.30	9.92
5	2.57	4.03
10	2.23	3.17
20	2.09	2.85
50	2.01	2.68
100	1.98	2.63

EXAMPLE 2.7.1

The sodium ion content of a urine specimen was determined by using an ion-selective electrode. The following values were obtained: 102, 97, 99, 98, 101, 106 mM. What are the 95% and 99% confidence limits for the sodium ion concentration?

The mean and standard deviation of these values are 100.5 mM and 3.27 mM respectively. There are six measurements and therefore 5 degrees of freedom. From Table A.2 the value of t_5 for calculating the 95% confidence limits is 2.57 and from equation (2.9) the 95% confidence limits of the mean are given by:

$$100.5 \pm 2.57 \times 3.27/\sqrt{6} = 100.5 \pm 3.4 \text{ mM}$$

Similarly the 99% confidence limits are given by:

$$100.5 \pm 4.03 \times 3.27/\sqrt{6} = 100.5 \pm 5.4 \text{ mM}$$

2.8 Presentation of results

As has already been emphasized, no quantitative experimental result is of any value unless it is accompanied by an estimate of the errors involved in its measurement. A common practice in analytical chemistry literature is to quote the mean as the estimate of the quantity measured and the standard deviation as the estimate of the precision. Less commonly, the standard error of the mean is sometimes quoted instead of the standard deviation, or the result is given in the form of the 95% confidence limits of the mean. (Uncertainty estimates, see Chapter 4, are also sometimes used.) Since there is no universal convention it is obviously essential to state the form used and, provided that the value of n is given, the three forms can be easily inter-converted by using equations (2.5) and (2.9).

A related aspect of presenting results is the rounding-off of the answer. The important principle here is that the number of significant figures given indicates the precision of the experiment. It would clearly be absurd, for example, to give the result of a titrimetric analysis as 0.107846 M – no analyst could achieve the implied precision of 0.000001 in ca. 0.1, i.e. 0.001%. In practice it is usual to quote as significant figures all the digits which are certain, plus the first uncertain one. For example, the mean of the values 10.09, 10.11, 10.09, 10.10 and 10.12 is 10.102, and their standard deviation is 0.01304. Clearly there is uncertainty in the second decimal place; the results are all 10.1 to one decimal place, but disagree in the second decimal place. Using the suggested method the result would be quoted as:

$$\bar{x} \pm s = 10.10 \pm 0.01 \quad (n = 5)$$

If it was felt that this resulted in an unacceptable rounding-off of the standard deviation, then the result could be given as:

$$\bar{x} \pm s = 10.10_2 \pm 0.01_3 \quad (n = 5)$$

where the use of a subscript indicates that the digit is given only to avoid loss of information. The reader could decide whether it was useful or not.

Similarly, when confidence limits are calculated [see equation (2.9)], there is no point in giving the value of $t_{n-1}s/\sqrt{n}$ to more than two significant figures. The value of $\bar{x}$ should then be given to the corresponding number of decimal places.

The number of significant figures quoted is sometimes used instead of a specific estimate of the precision of a result. For example 0.1046 M is taken to mean that the figures in the first three decimal places are certain but there is doubt about the fourth. Sometimes the uncertainty in the last figure is emphasized by using the formats 0.104(6) M or 0.104_6 M, but it remains preferable to give a specific estimate of precision such as the standard deviation.

One problem which arises is whether a 5 should be rounded up or down. For example, if 9.65 is rounded to one decimal place, should it become 9.6 or 9.7? It is evident that the results will be biased if a 5 is always rounded up; this bias can be avoided by rounding the 5 to the nearest even number giving, in this case, 9.6. Analogously, 4.75 is rounded to 4.8.

When several measured quantities are to be used to calculate a final result (see Section 2.11) these quantities should not be rounded-off too much or a needless loss of precision will result. A good rule is to keep one digit beyond the last significant figure and leave further rounding until the final result is reached. The same advice applies when the mean and standard deviation are used to apply a statistical test such as the *F*- and *t*-tests (see Chapter 3): the unrounded values of $\bar{x}$ and s should be used in the calculations.

2.9 Other uses of confidence limits

Confidence intervals can be used as a test for systematic errors as shown in the following example.

EXAMPLE 2.9.1

The absorbance scale of a spectrometer is tested at a particular wavelength with a standard solution which has an absorbance given as 0.470. Ten measurements of the absorbance with the spectrometer give $\bar{x} = 0.461$, and $s = 0.003$. Find the 95% confidence interval for the mean absorbance as measured by the spectrometer, and hence decide whether a systematic error is present.

The 95% confidence limits for the absorbance as measured by the spectrometer are [equation (2.9)]:

$$\bar{x} \pm t_{n-1}s/\sqrt{n} = 0.461 \pm 2.26 \times 0.003/\sqrt{10} = 0.461 \pm 0.002$$

(The value of t_9 was obtained from Table A.2).

Since the confidence interval does not include the known absorbance of 0.470, it is likely that a systematic error has occurred.

In practice the type of problem in Example 2.9.1 is usually tackled by a different, but related, approach (see Example 3.2.1).

Confidence limits can also be used in cases where measurements are made on each of a number of specimens. Suppose for example that the mean weight of a tablet in a very large batch is required: it would be too time-consuming to weigh each tablet. Similarly, if the mean iron content is measured using a destructive method of analysis such as atomic-absorption spectrometry, it is clearly impossible to examine every tablet. In each case, a sample could be taken from the batch (which in such instances forms the population), and from the mean and standard deviation of the sample a confidence interval could be found for the mean value of the quantity measured.

2.10 Confidence limits of the geometric mean for a log-normal distribution

In Section 2.3 it was mentioned that measurements on a number of different specimens may not be normally distributed. If they come from a log-normal distribution, then the confidence limits should be calculated taking this fact into account. Since the log of the measurements is normally distributed it is more accurate to work with the logarithms of the measurements when calculating a confidence interval. The confidence interval obtained will be the confidence interval for the *geometric* mean.

EXAMPLE 2.10.1

The following values (expressed as percentages) give the antibody concentration in human blood serum for a sample of eight healthy adults.

2.15, 1.13, 2.04, 1.45, 1.35, 1.09, 0.99, 2.07

Calculate the 95% confidence interval for the geometric mean assuming that the antibody concentration is log-normally distributed.

The logarithms (to the base 10) of these values are:

0.332, 0.053, 0.310, 0.161, 0.130, 0.037, −0.004, 0.316

The mean of these logged values is 0.1669, giving $10^{0.1669} = 1.47$ as the geometric mean of the original values. The standard deviation of the logged values is 0.1365.

The 95% confidence limits for the logged values are:

$$0.1669 \pm 2.36 \times 0.1365/\sqrt{8} = 0.1669 \pm 0.1139 = 0.0530 \text{ to } 0.2808$$

Taking antilogarithms of these limits gives the 95% confidence interval of the geometric mean as 1.13 to 1.91.

2.11 Propagation of random errors

In experimental work, the quantity to be determined is often calculated from a combination of observable quantities. We have already seen, for example, that even a relatively simple operation such as a titrimetric analysis involves several stages, each of which will be subject to errors (see Chapter 1). The final calculation may involve taking the sum, difference, product or quotient of two or more quantities or the raising of any quantity to a power.

It is most important to note that the procedures used for combining random and systematic errors are completely different. This is because random errors to some extent cancel each other out, whereas every systematic error occurs in a definite and known sense. Suppose for example that the final result of an experiment, x, is given by $x = a + b$. If a and b each have a systematic error of $+1$, it is clear that the systematic error in x is $+2$. If, however, a and b each have a random error of ± 1, the random error in x is not ± 2: this is because there will be occasions when the random error in a is positive while that in b is negative (or vice versa).

This section deals only with the propagation of random errors (systematic errors are considered in Section 2.12). If the precision of each observation is known then simple mathematical rules can be used to estimate the precision of the final result. These rules can be summarized as follows.

2.11.1 Linear combinations

In this case the final value, y, is calculated from a linear combination of measured quantities a, b, c, etc. by:

$$y = k + k_a a + k_b b + k_c c + \ldots \qquad (2.10)$$

where k, k_a, k_b, k_c, etc. are constants. Variance (defined as the square of the standard deviation) has the important property that the variance of a sum or difference of independent quantities is equal to the sum of their variances. It can be shown that if σ_a, σ_b, σ_c, etc. are the standard deviations of *a*, *b*, *c*, etc., then the standard deviation of *y*, σ_y, is given by:

$$\sigma_y = \sqrt{(k_a\sigma_a)^2 + (k_b\sigma_b)^2 + (k_c\sigma_c)^2 + \ldots} \tag{2.11}$$

EXAMPLE 2.11.1

In a titration the initial reading on the burette is 3.51 ml and the final reading is 15.67 ml, both with a standard deviation of 0.02 ml. What is the volume of titrant used and what is its standard deviation?

$$\text{Volume used} = 15.67 - 3.51 = 12.16 \text{ ml}$$

$$\text{Standard deviation} = \sqrt{(0.02)^2 + (0.02)^2} = 0.028 \text{ ml}$$

This example illustrates the important point that the standard deviation for the final result is larger than the standard deviations of the individual burette readings, even though the volume used is calculated from a difference. It is, however, less than the sum of the standard deviations.

2.11.2 Multiplicative expressions

If *y* is calculated from an expression of the type:

$$y = kab/cd \tag{2.12}$$

(where *a*, *b*, *c* and *d* are independent measured quantities and *k* is a constant) then there is a relationship between the squares of the *relative* standard deviations:

$$\frac{\sigma_y}{y} = \sqrt{\left(\frac{\sigma_a}{a}\right)^2 + \left(\frac{\sigma_b}{b}\right)^2 + \left(\frac{\sigma_c}{c}\right)^2 + \left(\frac{\sigma_d}{d}\right)^2} \tag{2.13}$$

EXAMPLE 2.11.2

The quantum yield of fluorescence, ϕ, is calculated from the expression:

$$\phi = I_f/kclI_0\varepsilon$$

where the quantities involved are defined below, with an estimate of their relative standard deviations in brackets:

I_0 = incident light intensity (0.5%)
I_f = fluorescence intensity (2%)
ε = molar absorptivity (1%)
c = concentration (0.2%)
l = path-length (0.2%)
k is an instrument constant.

From equation (2.13), the relative standard deviation of ϕ is given by:

$$\text{RSD} = \sqrt{2^2 + 0.2^2 + 0.2^2 + 0.5^2 + 1^2} = 2.3\%$$

It can be seen that the relative standard deviation in the final result is not much larger than the largest relative standard deviation used to calculate it (i.e. 2% for I_f). This is mainly a consequence of the squaring of the relative standard deviations and illustrates an important general point: any efforts to improve the precision of an experiment need to be directed towards improving the precision of the least precise values. As a corollary to this, there is no point in wasting effort in increasing the precision of the most precise values. This is not to say that small errors are unimportant: small errors at many stages of an experiment, such as the titrimetric analysis discussed in detail in Chapter 1, will produce an appreciable error in the final result.

It is important to note that when a quantity is raised to a power, e.g. b^3, then the error is not calculated as for a multiplication, i.e. $b \times b \times b$, because the quantities involved are not independent. If the relationship is:

$$y = b^n \tag{2.14}$$

then the standard deviations of y and b are related by:

$$\left|\frac{\sigma_y}{y}\right| = \left|\frac{n\sigma_b}{b}\right| \tag{2.15}$$

(The modulus sign means that the magnitude of the enclosed quantity is taken without respect to sign, e.g. $|-2| = 2$.)

2.11.3 Other functions

If y is a general function of x, $y = f(x)$, then the standard deviations of x and y are related by:

$$\sigma_y = \left|\sigma_x \frac{dy}{dx}\right| \tag{2.16}$$

EXAMPLE 2.11.3

The absorbance, A, of a solution is given by $A = -\log(T)$ where T is the transmittance. If the measured value of T is 0.501 with a standard deviation of 0.001, calculate A and its standard deviation.

We have:

$$A = -\log 0.501 = 0.300$$

Also:

$$\mathrm{d}A/\mathrm{d}T = -(\log e)/T = -0.434/T$$

so from equation (2.17):

$$\sigma_A = |\sigma_T(-\log e/T)| = |0.001 \times (-0.434/0.501)| = 0.0008_7$$

It is interesting to note that for this widely used experimental method we can also find the conditions for which the relative standard deviation is a minimum. The relative standard deviation (r.s.d.) of A is given by:

$$\text{r.s.d. of } A = 100\sigma_A/A = \frac{100\sigma_T \log e}{T \log T}$$

Differentiation of this expression with respect to T shows that the r.s.d. of A is a minimum when $T = 1/e = 0.368$.

2.12 Propagation of systematic errors

The rules for the combination of systematic errors can also be divided into three groups.

2.12.1 Linear combinations

If y is calculated from measured quantities by use of equation (2.10), and the systematic errors in a, b, c, etc., are Δa, Δb, Δc, etc., then the systematic error in y, Δy, is calculated from:

$$\Delta y = k_a\Delta a + k_b\Delta b + k_c\Delta c + \ldots \qquad (2.17)$$

Remember that the systematic errors are either positive or negative and that these signs must be included in the calculation of Δy.

The total systematic error can sometimes be zero. Suppose, for example, a balance with a systematic error of −0.01 g is used for the weighings involved in making a standard solution. Since the weight of the solute used is found from the difference between two weighings, the systematic errors cancel out. It should be pointed out that this applies only to an electronic balance with a single internal reference weight. Carefully considered procedures, such as this, can often minimize the systematic errors, as described in Chapter 1.

2.12.2 Multiplicative expressions

If y is calculated from the measured quantities by use of equation (2.12) then *relative* systematic errors are used:

$$(\Delta y/y) = (\Delta a/a) + (\Delta b/b) + (\Delta c/c) + (\Delta d/d) \tag{2.18}$$

When a quantity is raised to some power, then equation (2.15) is used with the modulus sign omitted and the standard deviations replaced by systematic errors.

2.12.3 Other functions

The equation used is identical to equation (2.16) but with the modulus sign omitted and the standard deviations replaced by systematic errors.

In any real analytical experiment both random and systematic errors will occur. The estimated combined error in the final result is now referred to as the **uncertainty**. The uncertainty combines random and systematic errors and provides a realistic range of values within which the true value of a measured quantity probably lies. This topic is dealt with in detail in Chapter 4.

Bibliography

Altman, D. G. 1991. *Practical Statistics for Medical Research*. Chapman and Hall, London. (Gives a fuller discussion of the log-normal distribution.)

Commission on Analytical Nomenclature. 1990. Nomenclature for Sampling in Analytical Chemistry. *Pure and Applied Chemistry* 62: 1193.

Davies, O. L. and Goldsmith, P. L. 1982. *Statistical Methods in Research and Production*. Longman, London. (Gives a more detailed treatment of the subject matter of this chapter.)

Moritz, P. 1981. Chapter 1 in *Comprehensive Analytical Chemistry Vol. XI*, G. Svehla (ed.). Elsevier, Amsterdam. (This article discusses in detail the application of the theory of errors to analytical methods.)

Skoog, D. A. and West, D. M. 1982. *Fundamentals of Analytical Chemistry*, 4th Ed. Holt Saunders, New York. (Describes the use of statistics in evaluating analytical data.)

Exercises

1. The reproducibility of a method for the determination of selenium in foods was investigated by taking nine samples from a single batch of brown rice and determining the selenium concentration in each. The following results were obtained:

   0.07 0.07 0.08 0.07 0.07 0.08 0.08 0.09 0.08 $\mu g\,g^{-1}$

   (Moreno-Domínguez, T., García-Moreno, C. and Mariné-Font, A. 1983. *Analyst* 108: 505)

   Calculate the mean, standard deviation and relative standard deviation of these results.

2. Seven measurements of the pH of a buffer solution gave the following results:

   5.12 5.20 5.15 5.17 5.16 5.19 5.15

   Calculate (i) the 95% and (ii) the 99% confidence limits for the true pH. (Assume that there are no systematic errors.)

3. Ten replicate analyses of the concentration of mercury in a sample of commercial gas condensate gave the following results:

   23.3 22.5 21.9 21.5 19.9 21.3 21.7 23.8 22.6 24.7 $ng\,ml^{-1}$

   (Shafawi, A., Ebdon, L., Foulkes, M., Stockwell, P. and Corns, W. 1999. *Analyst* 124: 185)

   Calculate the mean, standard deviation, relative standard deviation and 99% confidence limits of the mean.

   Six replicate analyses on another sample gave the following values:

   13.8 14.0 13.2 11.9 12.0 12.1 $ng\,ml^{-1}$

   Repeat the calculations for these values.

4. The concentration of lead in the bloodstream was measured for a sample of 50 children from a large school near a busy main road. The sample mean was 10.12 $ng\,ml^{-1}$ and the standard deviation was 0.64 $ng\,ml^{-1}$. Calculate the 95% confidence interval for the mean lead concentration for all the children in the school.

   About how big should the sample have been to reduce the range of the confidence interval to 0.2 $ng\,ml^{-1}$ (i.e. $\pm$0.1 $ng\,ml^{-1}$)?

5. In an evaluation of a method for the determination of fluorene in sea-water, a synthetic sample of sea-water was spiked with 50 $ng\,ml^{-1}$ of fluorene. Ten replicate determinations of the fluorene concentration in the sample had a mean of 49.5 $ng\,ml^{-1}$ with a standard deviation of 1.5 $ng\,ml^{-1}$.

   (Gonsález, M. A. and López, M. H. 1998. *Analyst* 123: 2217)

Calculate the 95% confidence limits of the mean. Is the spiked value of 50 ng ml^{-1} within the 95% confidence limits?

6. A 0.1 M solution of acid was used to titrate 10 ml of 0.1 M solution of alkali and the following volumes of acid were recorded:

9.88 10.18 10.23 10.39 10.21 ml

Calculate the 95% confidence limits of the mean and use them to decide whether there is any evidence of systematic error.

7. This problem considers the random errors involved in making up a standard solution. A volume of 250 ml of a 0.05 M solution of a reagent of formula weight (relative molecular mass) 40 was made up, the weighing being done by difference. The standard deviation of each weighing was 0.0001 g: what were the standard deviation and relative standard deviation of the weight of reagent used? The standard deviation of the volume of solvent used was 0.05 ml. Express this as a relative standard deviation. Hence calculate the relative standard deviation of the molarity of the solution.

Repeat the calculation for a reagent of formula weight 392.

8. The solubility product of barium sulphate is 1.3×10^{-10}, with a standard deviation of 0.1×10^{-10}. Calculate the standard deviation of the calculated solubility of barium sulphate in water.

Significance tests

3.1 Introduction

One of the most important properties of an analytical method is that it should be free from systematic error. This means that the value which it gives for the amount of the analyte should be the true value. This property of an analytical method may be tested by applying the method to a standard test portion containing a known amount of analyte (Chapter 1). However, as we saw in the last chapter, even if there were no systematic error, random errors make it most unlikely that the measured amount would exactly equal the standard amount. In order to decide whether the difference between the measured and standard amounts can be accounted for by random error a statistical test known as a **significance test** can be employed. As its name implies, this approach tests whether the difference between the two results is significant, or whether it can be accounted for merely by random variations. Significance tests are widely used in the evaluation of experimental results. This chapter considers several tests which are particularly useful to analytical chemists.

3.2 Comparison of an experimental mean with a known value

In making a significance test we are testing the truth of a hypothesis which is known as a **null hypothesis**, often denoted by $\mathbf{H_0}$. For the example in the previous paragraph we adopt the null hypothesis that the analytical method is not subject to systematic error. The term *null* is used to imply that there is no difference between the observed and known values other than that which can be attributed to random variation. Assuming that this null hypothesis is true, statistical theory can be used to calculate the probability that the observed difference (or a greater one) between the sample mean, $\bar{x}$, and the true value, μ, arises solely as a result of random errors. The lower the probability that the observed difference occurs by chance, the less likely it is that the null hypothesis is true. Usually the null hypothesis is rejected if the probability of such a difference occurring by chance is less than 1 in 20 (i.e. 0.05 or 5%). In such a case the difference is said to be **significant at the 0.05 (or 5%) level**. Using this level of significance there is, on average, a 1 in 20 chance that we shall reject the null hypothesis *when it is in fact true*. In order to be more certain that we make the correct decision a higher level of significance can be used, usually 0.01 or 0.001 (1% or 0.1%). The significance level is indicated by writing, for example, P (i.e. probability) $= 0.05$, and gives the probability of rejecting a true null hypothesis. It is important to appreciate that if the null hypothesis is retained it has not been *proved* that it is true, only that it has not been demonstrated to be false. Later in the chapter the probability of retaining a null hypothesis when it is in fact false will be discussed.

In order to decide whether the difference between $\bar{x}$ and μ is significant, that is to test H_0: population mean $= \mu$, the statistic t is calculated:

$$t = (\bar{x} - \mu)\sqrt{n}/s \tag{3.1}$$

where $\bar{x}$ = sample mean, s = sample standard deviation and n = sample size.

If $|t|$ (i.e. the calculated value of t without regard to sign) exceeds a certain **critical value** then the null hypothesis is rejected. The critical value of t for a particular significance level can be found from Table A.2. For example, for a sample size of 10 (i.e. 9 degrees of freedom) and a significance level of 0.01, the critical value is $t_9 = 3.25$, where, as in Chapter 2, the subscript is used to denote the number of degrees of freedom.

EXAMPLE 3.2.1

In a new method for determining selenourea in water, the following values were obtained for tap water samples spiked with 50 ng ml^{-1} of selenourea:

50.4,   50.7,   49.1,   49.0,   51.1 ng ml^{-1}

(Aller, A. J. and Robles, L. C. 1998. *Analyst* 123: 919).

Is there any evidence of systematic error?

The mean of these values is 50.06 and the standard deviation is 0.956. Adopting the null hypothesis that there is no systematic error, i.e. $\mu = 50$, and using equation

(3.1) gives

$$t = \frac{(50.06 - 50)\sqrt{5}}{0.956} = 0.14$$

From Table A.2, the critical value is $t_4 = 2.78$ ($P = 0.05$). Since the observed value of $|t|$ is less than the critical value the null hypothesis is retained: there is no evidence of systematic error. Note again that this does not mean that there are no systematic errors, only that they have not been demonstrated.

The use in significance testing of critical values from statistical tables was adopted because it was formerly too tedious to calculate the probability of t exceeding the experimental value. Computers have altered this situation, and statistical software usually quotes the results of significance tests in terms of a probability. If the individual data values are entered in Minitab the result of performing this test is shown below.

```
t-Test of the mean

Test of mu = 50.000 vs mu not = 50.000

Variable N Mean    StDev SE Mean T     P
selenour 5 50.060 0.956 0.427   0.14 0.90
```

This gives the additional information that $P\ (|t| > 0.14) = 0.90$. Since this probability is much greater than 0.05, the result is not significant at $P = 0.05$, in agreement with the previous calculation. Obviously the power to calculate an exact probability is a great advantage, removing the need for statistical tables containing critical values. The examples in this book, however, use critical values, as not all readers may have access to suitable software, and many scientists continue to perform significance tests with the aid of hand-held calculators, which do not normally provide P values. Also in cases where the individual data values are not provided, Minitab or Excel, for example, cannot be used. However, where the calculation can be performed using these programs, the P value will also be quoted.

3.3 Comparison of two experimental means

Another way in which the results of a new analytical method may be tested is by comparing them with those obtained by using a second (perhaps a reference) method. In this case we have two sample means $\bar{x}_1$ and $\bar{x}_2$. Taking the null hypothesis that the two methods give the same result, that is H_0: $\mu_1 = \mu_2$, we need to test whether $(\bar{x}_1 - \bar{x}_2)$ differs significantly from zero. If the two samples have standard deviations which are not significantly different (see Section 3.5 for a method of testing this assumption), a pooled estimate, s, of the standard deviation can be calculated from the two individual standard deviations s_1 and s_2.

In order to decide whether the difference between two sample means, $\bar{x}_1$ and $\bar{x}_2$ is significant, that is to test the null hypothesis, H_0: $\mu_1 = \mu_2$, the statistic t is calculated:

$$t = \frac{(\bar{x}_1 - \bar{x}_2)}{s\sqrt{\dfrac{1}{n_1} + \dfrac{1}{n_2}}} \tag{3.2}$$

where s is calculated from:

$$s^2 = \frac{(n_1 - 1)s_1^2 + (n_2 - 1)s_2^2}{(n_1 + n_2 - 2)} \tag{3.3}$$

and t has $n_1 + n_2 - 2$ degrees of freedom.

This method assumes that the samples are drawn from populations with equal standard deviations.

EXAMPLE 3.3.1

In a comparison of two methods for the determination of chromium in rye grass, the following results (mg kg^{-1} Cr) were obtained:

Method 1: mean = 1.48; standard deviation 0.28
Method 2: mean = 2.33; standard deviation 0.31

For each method five determinations were made.
(Sahuquillo, A., Rubio, R. and Rauret, G. 1999. *Analyst* 124: 1)

Do these two methods give results having means which differ significantly?

The null hypothesis adopted is that the means of the results given by the two methods are equal. From equation (3.3), the pooled value of the standard deviation is given by:

$$s^2 = ([4 \times 0.28^2] + [4 \times 0.31^2])/(5 + 5 + 2) = 0.0873$$

$$s = 0.295$$

From equation (3.2):

$$t = \frac{2.33 - 1.48}{0.295\sqrt{\frac{1}{5} + \frac{1}{5}}} = 4.56$$

There are 8 degrees of freedom, so (Table A.2) the critical value, $t_8 = 2.31$ ($P = 0.05$): since the experimental value of $|t|$ is greater than this the difference between the two results is significant at the 5% level and the null hypothesis is rejected. In fact since the critical value of t for $P = 0.01$ is about 3.36, the difference is significant at the 1% level. In other words, if the null hypothesis is true the probability of such a large difference arising by chance is less than 1 in 100.

Another application of this test is illustrated by the following example where it is used to decide whether a change in the conditions of an experiment affects the result.

EXAMPLE 3.3.2

In a series of experiments on the determination of tin in foodstuffs, samples were boiled with hydrochloric acid under reflux for different times. Some of the results are shown below:

Refluxing time (min)	Tin found (mg kg^{-1})
30	55, 57, 59, 56, 56, 59
75	57, 55, 58, 59, 59, 59

(Analytical Methods Committee. 1983. *Analyst* 108: 109)

Does the mean amount of tin found differ significantly for the two boiling times?

The mean and variance (square of the standard deviation) for the two times are:

$$30 \text{ min} \quad \bar{x}_1 = 57.00 \quad s_1^2 = 2.80$$

$$75 \text{ min} \quad \bar{x}_2 = 57.83 \quad s_2^2 = 2.57$$

The null hypothesis is adopted that boiling has no effect on the amount of tin found. By equation (3.3), the pooled value for the variance is given by:

$$s^2 = (5 \times 2.80 + 5 \times 2.57)/10 = 2.685$$

$$s = 1.64$$

From equation (3.2):

$$t = \frac{57.00 - 57.83}{1.64\sqrt{\frac{1}{6} + \frac{1}{6}}} = -0.88$$

$$= -\,0.88$$

There are 10 degrees of freedom so the critical value is $t_{10} = 2.23$ ($P = 0.05$). The observed value of $|t|$ (=0.88) is less than the critical value so the null hypothesis is retained: there is no evidence that the length of boiling time affects the recovery rate.

The table below shows the result of performing this calculation using Excel.

t-Test: two-sample assuming equal variances

	Variable 1	Variable 2
Mean	57	57.833
Variance	2.8	2.567
Observations	6	6
Pooled variance	2.683	
Hypothesized mean difference	0	
df	10	
t Stat	−0.881	
P(T<=t) one-tail	0.199	
t Critical one-tail	1.812	
P(T<=t) two-tail	0.399	
t Critical two-tail	2.228	

The distinction between 'one-tail' and 'two-tail' will be covered in Section 3.5. For the present, it is sufficient to consider only the two-tail values. These show that $P(|t| > 0.88) = 0.399$. Since this probability is greater than 0.05, the result is not significant at the 5% level.

If the population standard deviations are unlikely to be equal then it is no longer appropriate to pool sample standard deviations in order to give an overall estimate of standard deviation. An approximate method in these circumstances is given below:

In order to test H_0: $\mu_1 = \mu_2$ when it cannot be assumed that the two samples come from populations with equal standard deviations, the statistic t is calculated where

$$t = \frac{(\bar{x}_1 - \bar{x}_2)}{\sqrt{\dfrac{s_1^2}{n_1} + \dfrac{s_2^2}{n_2}}} \tag{3.4}$$

$$\text{with degrees of freedom} = \frac{\left(\dfrac{s_1^2}{n_1} + \dfrac{s_2^2}{n_2}\right)^2}{\left(\dfrac{s_1^4}{n_1^2(n_1 - 1)} + \dfrac{s_2^4}{n_2^2(n_2 - 1)}\right)} \tag{3.5}$$

with the value obtained being truncated to an integer.

The reader should be aware that there are various versions given in the literature for the number of degrees of freedom for t, reflecting the fact that the method is an approximate one. The method above is that used by Minitab and it errs on the side of caution in giving a significant result. Excel, on the other hand, uses equation (3.5) but rounds the value to the nearest integer. For example, if equation (3.5) gave a value of 4.7, Minitab would take 4 degrees of freedom and Excel would take 5.

EXAMPLE 3.3.3

The data below give the concentration of thiol (mM) in the blood lysate of the blood of two groups of volunteers, the first group being 'normal' and the second suffering from rheumatoid arthritis:

Normal:	1.84,	1.92,	1.94,	1.92,	1.85,	1.91,	2.07
Rheumatoid:	2.81,	4.06,	3.62,	3.27,	3.27,	3.76	

(Banford, J. C., Brown, D. H., McConnell, A. A., McNeil, C. J., Smith, W. E., Hazelton, R. A. and Sturrock, R. D. 1983. *Analyst* 107: 195)

The null hypothesis adopted is that the mean concentration of thiol is the same for the two groups.

The reader can check that:

$$n_1 = 7 \qquad \bar{x}_1 = 1.921 \qquad s_1 = 0.076$$
$$n_2 = 6 \qquad \bar{x}_2 = 3.465 \qquad s_2 = 0.440$$

Substitution in equation (3.4) gives $t = -8.48$ and substitution in equation (3.5) gives 5.3, which is truncated to 5. The critical value is $t_5 = 4.03$ ($P = 0.01$) so the null hypothesis is rejected: there is sufficient evidence to say that the mean concentration of thiol differs between the groups.

The result of performing this calculation using Minitab (where the non-pooled test is the default option) is shown below.

Two sample t-test and confidence interval

```
Two sample T for Normal vs Rheumatoid

            N    Mean      StDev     SE Mean
Normal      7    1.9214    0.0756    0.029
Rheumato    6    3.465     0.440     0.18

95% CI for mu Normal - mu Rheumato: (-2.012, -1.08)
T-Test mu Normal = mu Rheumato (vs not =): T= -8.48
P = 0.0004 DF = 5
```

This confirms the values above and also gives the information that $P(|t| > 8.48) = 0.0004$. This probability is extremely low: the result is in fact significant at $P = 0.001$.

3.4 Paired *t*-test

It frequently happens that two methods of analysis are compared by studying test samples containing different amounts of analyte. For example, Table 3.1 gives the results of determining paracetamol concentration (% m/m) in tablets by two different methods. Ten tablets from ten different batches were analysed in order to see whether the results obtained by the two methods differed.

As always there is the variation between the measurements due to random measurement error. In addition, differences between the tablets and differences between the methods may also contribute to the variation between measurements. It is the latter which is of interest in this example: we wish to know whether the methods produce significantly different results. The test for comparing two means (Section 3.3) is not appropriate in this case because it does not separate the variation due to method from that due to variation between tablets: the two effects are said to be 'confounded'. This difficulty is overcome by looking at the difference, *d*, between each pair of results given by the two methods. If there is no difference between the two methods then these differences are drawn from a population with mean

Table 3.1 Example of paired data

Batch	*UV spectrometric assay*	*Near-infrared reflectance spectroscopy*
1	84.63	83.15
2	84.38	83.72
3	84.08	83.84
4	84.41	84.20
5	83.82	83.92
6	83.55	84.16
7	83.92	84.02
8	83.69	83.60
9	84.06	84.13
10	84.03	84.24

(Trafford, A. D., Jee, R. D., Moffat, A. C. and Graham, P. 1999. *Analyst* 124: 163)

$\mu_d = 0$. In order to test the null hypothesis, we test whether $\bar{d}$ differs significantly from 0 using the statistic t.

To test whether n paired results are drawn from the same population, that is H_0: $\mu_d = 0$, calculate the statistic t:

$$t = \bar{d}\sqrt{n}/s_d \tag{3.6}$$

where $\bar{d}$ and s_d are the mean and standard deviation respectively of d, the difference between paired values.

The number of degrees of freedom of t is $n - 1$.

EXAMPLE 3.4.1

Test whether there is a significant difference between the results obtained by the two methods in Table 3.1.

The differences between the pairs of values (taking the second value from the first value) are:

+1.48, +0.66, +0.24, +0.21, −0.10, −0.61, −0.10, +0.09, −0.07, −0.21

These values have mean, $\bar{d} = 0.159$ and standard deviation, $s_d = 0.570$. Substituting in equation (3.6), with $n = 10$, gives $t = 0.88$. The critical value is $t_9 = 2.26$ ($P = 0.05$). Since the calculated value of $|t|$ is less than this the null hypothesis is not rejected: the methods do not give significantly different results for the paracetamol concentration.

Again this calculation can be performed on a computer, giving the result that $P(|t| \geq 0.882) = 0.40$. Since this probability is greater than 0.05 we reach the same conclusion: the two methods do not differ significantly at $P = 0.05$.

The paired test described above does not require that the precisions of the two methods are equal but it does assume that the differences, d, are normally distributed. This will be the case if each method's measurement is normally distributed and the precision and bias (if any) of each method are constant over the range of values for which the measurements were made. The data can consist of either single measurements, as in Example 3.4.1, or the means of replicate measurements. However, it is necessary for the same number of measurements to be made on each sample by the first method and likewise for the second method: that is n measurements on each sample by method 1 and m measurements on each sample by method 2, where m and n do not have to be equal.

There are various circumstances in which it may be necessary or desirable to design an experiment so that each sample is analysed by each of two methods, giving results that are naturally paired. Some examples are:

1. The quantity of any one test sample is sufficient for only one determination by each method.
2. The test samples may be presented over an extended period so it is necessary to remove the effects of variations in the environmental conditions such as temperature, pressure, etc.
3. The methods are to be compared by using a wide variety of samples from different sources and possibly with very different concentrations (but see the next paragraph).

As analytical methods usually have to be applicable over a wide range of concentrations, a new method is often compared with a standard method by analysis of samples in which the analyte concentration may vary over several powers of ten. In this case it is inappropriate to use the paired t-test since its validity rests on the assumption that any errors, either random or systematic, are independent of concentration. Over wide ranges of concentration this assumption may no longer be true. An alternative method in such cases is linear regression (see Section 5.9) but this approach also presents difficulties.

3.5 One-sided and two-sided tests

The methods described so far in this chapter have been concerned with testing for a difference between two means in either direction. For example, the method described in Section 3.2 tests whether there is a significant difference between the experimental result and the known value for the reference material, regardless of the sign of the difference. In most situations of this kind the analyst has no idea, prior to the experiment, as to whether any difference between the experimental mean and the reference value will be positive or negative. Thus the test used must cover either possibility. Such a test is called **two-sided** (or two-tailed). In a few cases, however, a different kind of test may be appropriate. Consider, for example, an experiment in which it is hoped to increase the rate of reaction by addition of a catalyst. In this case, it is clear before the experiment begins that the only result of interest is whether the new rate is greater than the old, and only an increase need be tested for significance. This kind of test is called **one-sided** (or one-tailed). For a given value of n and a particular probability level, the critical value for a one-sided test differs from that for a two-sided test. In a one-sided test for an increase, the critical value of t (rather than $|t|$) for $P = 0.05$ is that value which is exceeded with a probability of 5%. Since the sampling distribution of the mean is assumed to be symmetrical, this probability is twice the probability that is relevant in the two-sided test. The appropriate value for the one-sided test is thus found in the $P = 0.10$ column of Table A.2. Similarly, for a one-sided test at the $P = 0.01$ level, the 0.02 column is used. For a one-sided test for a decrease, the critical value of t will be of equal magnitude but with a negative sign. If the test is carried out on a computer, it will be necessary to indicate whether a one- or a two-sided test is required.

EXAMPLE 3.5.1

It is suspected that an acid–base titrimetric method has a significant indicator error and thus tends to give results with a positive systematic error (i.e. positive bias). To test this an exactly 0.1 M solution of acid is used to titrate 25.00 ml of an exactly 0.1 M solution of alkali, with the following results (ml):

25.06 25.18 24.87 25.51 25.34 25.41

Test for positive bias in these results.

For these data we have:

$$\text{mean} = 25.228 \text{ ml, standard deviation} = 0.238 \text{ ml}$$

Adopting the null hypothesis that there is no bias, H_0: $\mu = 25.00$, and using equation (3.1) gives:

$$t = (25.228 - 25.00) \times \sqrt{6}/0.238 = 2.35$$

From Table A.2 the critical value is $t_5 = 2.02$ ($P = 0.05$, one-sided test). Since the observed value of t is greater than this the null hypothesis is rejected and there is evidence for positive bias.

Using a computer gives $P(t \geq 2.35) = 0.033$. Since this is less than 0.05, the result is significant at $P = 0.05$, as before.

It is interesting to note that if a two-sided test had been made in the example above (for which the critical value for $t_5 = 2.57$) the null hypothesis would not have been rejected! This apparently contradictory result is explained by the fact that the decision on whether to make a one- or two-sided test depends on the degree of prior knowledge, in this case a suspicion or expectation of positive bias. Obviously it is essential that the decision on whether the test is one- or two-sided should be made before the experiment has been done, and not with hindsight, when the results might prejudice the choice. In general, it will be found that two-sided tests are much more commonly used than one-sided ones. The relatively rare circumstances in which one-sided tests are necessary are easily identified.

3.6 *F*-test for the comparison of standard deviations

The significance tests described so far are used for comparing means, and hence for detecting systematic errors. In many cases it is also important to compare the standard deviations, i.e. the random errors of two sets of data. As with tests on means, this comparison can take two forms. Either we may wish to test whether Method A is more precise than Method B (i.e. a one-sided test) or we may wish to test whether Methods A and B differ in their precision (i.e. a two-sided test). For example, if we wished to test whether a new analytical method is more precise than a standard method we would use a one-sided test; if we wished to test whether two standard deviations differ significantly (e.g. before applying a *t*-test – see Section 3.3 above) a two-sided test is appropriate.

The *F*-test considers the ratio of the two sample variances, i.e. the ratio of the squares of the standard deviations, s_1^2/s_2^2.

In order to test whether the difference between two sample variances is significant, that is to test H_0: $\sigma_1^2 = \sigma_2^2$, the statistic F is calculated:

$$F = s_1^2/s_2^2 \qquad (3.7)$$

where 1 and 2 are allocated in the equation so that F is always ≥ 1.

The number of degrees of freedom of the numerator and denominator are $n_1 - 1$ and $n_2 - 1$ respectively.

The test assumes that the populations from which the samples are taken are normal.

If the null hypothesis is true then the variance ratio should be close to 1. Differences from 1 can occur because of random variation but if the difference is too great it can no longer be attributed to this cause. If the calculated value of F exceeds a certain critical value (obtained from tables) then the null hypothesis is rejected. This critical value of F depends on the size of both the samples, the significance level and the type of test performed. The values for $P = 0.05$ are given in Appendix 2 in Table A.3 for one-sided tests and Table A.4 for two-sided tests; the use of these tables is illustrated in the following examples.

EXAMPLE 3.6.1

A proposed method for the determination of the chemical oxygen demand of wastewater was compared with the standard (mercury salt) method. The following results were obtained for a sewage effluent sample:

	Mean ($mg\,l^{-1}$)	*Standard deviation* ($mg\,l^{-1}$)
Standard method	72	3.31
Proposed method	72	1.51

For each method eight determinations were made.
(Ballinger, D., Lloyd, A. and Morrish, A. 1982. *Analyst* 107: 1047)

Is the precision of the proposed method significantly greater than that of the standard method?

We have to decide whether the variance of the standard method is significantly greater than that of the proposed method. F is given by the ratio of the variances:

$$F = \frac{3.31^2}{1.51^2} = 4.8$$

This is a case where a one-sided test must be used, the only point of interest being whether the proposed method is more precise than the standard method. In Table A.3 the number of degrees of freedom of the denominator is given in the left-hand column and the number of degrees of freedom of the numerator at the top. Both samples contain eight values so the number of degrees of freedom in each case is 7. The critical value is $F_{7,7} = 3.787$ ($P = 0.05$), where the subscripts indicate the degrees of freedom of the numerator and denominator respectively. Since the calculated value of F (4.8) exceeds this, the variance of the standard method is significantly greater than that of the proposed method at the 5% probability level, i.e. the proposed method is more precise.

EXAMPLE 3.6.2

In Example 3.3.1 it was assumed that the variances of the two methods for determining chromium in rye grass did not differ significantly. This assumption can now be tested. The standard deviations were 0.28 and 0.31 (each obtained from five measurements on a specimen of a particular plant). Calculating F so that it is greater than 1, we have:

$$F = \frac{0.31^2}{0.28^2} = 1.23$$

In this case, however, we have no reason to expect in advance that the variance of one method should be greater than the other, so a two-sided test is appropriate. The critical values given in Table A.3 are the values that F exceeds with a probability of 0.05 assuming that it must be greater than 1. In a two-sided test the ratio of the first to the second variance could be less or greater than 1, but if F is calculated so that it is greater than 1, the probability that it exceeds the critical values given in Table A.3 will be doubled. Thus these critical values are not appropriate for a two-sided test and Table A.4 is used instead. From this table, taking the number of degrees of freedom of both numerator and denominator as 4, the critical value is $F_{4,4} = 9.605$. The calculated value is less than this, so there is no significant difference between the two variances at the 5% level.

As with the t-test, other significance levels may be used for the F-test and the critical values can be found from the tables listed in the bibliography at the end of Chapter 1. Care must be taken that the correct table is used depending on whether the test is one- or two-sided: for an α% significance level the 2α% points of the F distribution are used for a one-sided test and the α% points are used for a two-sided test. If a computer is used it will be possible to obtain a P-value. Note that Excel only carries out a one-sided F-test and that it is necessary to enter the sample with the larger variance as the first sample. Minitab does not give an F-test for comparing the variances of two samples.

3.7 Outliers

Every experimentalist is familiar with the situation in which one (or possibly more) of a set of results appears to differ unreasonably from the others in the set. Such a measurement is called an outlier. In some cases an outlier may be attributed to a human error. For example if the following results were given for a titration:

12.12, 12.15, 12.13, 13.14, 12.12 ml

then the fourth value is almost certainly due to a slip in writing down the result and should read 12.14. However, even when such obviously erroneous values have been removed or corrected, values which appear to be outliers may still occur. Should they be kept, come what may, or should some means be found to test statistically whether or not they should be rejected? Obviously the final values presented for the mean and standard deviation will depend on whether or not the outliers are rejected. Since discussion of the precision and accuracy of a method depends on these final values, it should always be made clear whether outliers have been rejected, and if so, why.

Dixon's test (sometimes called the ***Q*-test**) is a popular test for outliers because the calculation is simple. For small samples (size 3 to 7) the test assesses a suspect measurement by comparing the difference between it and the measurement nearest to it in size with range of the measurements. (For larger samples the form of the test is modified slightly. A reference containing further details is given at the end of this chapter.)

In order to use Dixon's test for an outlier, that is to test H_0: all measurements come from the same population, the statistic Q is calculated:

$$Q = |\text{suspect value} - \text{nearest value}|/(\text{largest value} - \text{smallest value}) \quad (3.8)$$

This test is valid for samples size 3 to 7 and assumes that the population is normal.

The critical values of Q for $P = 0.05$ are given in Table A.5. If the calculated value of Q exceeds the critical value the suspect value is rejected. The values given are for a two-sided test, which is appropriate when it is not known in advance at which extreme an outlier may occur.

EXAMPLE 3.7.1

The following values were obtained for the nitrite concentration ($mg\,l^{-1}$) in a sample of river water:

0.403, 0.410, 0.401, 0.380

The last measurement is suspect: should it be rejected?

We have:

$$Q = |0.380 - 0.401|/(0.410 - 0.380) = 0.021/0.03 = 0.7$$

From Table A.5, for sample size 4, the critical value of Q is 0.831 ($P = 0.05$). Since the calculated value of Q does not exceed this the suspect measurement should be retained.

Ideally, further measurements should be made when a suspect value occurs, particularly if only a few values have been obtained initially. This may make it clearer whether or not the suspect value should be rejected, and, if it is still retained, will also reduce to some extent its effect on the mean and standard deviation.

EXAMPLE 3.7.2

If three further measurements were added to those given in the example above so that the complete results became:

0.403, 0.410, 0.401, 0.380, 0.400, 0.413, 0.411

should 0.380 still be retained?

The calculated value of Q is now:

$$Q = |0.380 - 0.400|/(0.413 - 0.380) = 0.606$$

The critical value of Q ($P = 0.05$) for a sample size 7 is 0.570 so the suspect measurement is rejected at the 5% significance level.

Another frequently used test for outliers is **Grubbs' test**, which compares the deviation of the suspect value from the sample mean with the standard deviation of the sample. This test is recommended by ISO in preference to Dixon's test.

In order to use Grubbs' test for an outlier, that is to test H_0: all measurements come from the same population, the statistic G is calculated:

$$G = |\text{suspect value} - \bar{x}|/s \qquad (3.9)$$

where s is calculated with the suspect value included.

The test assumes that the population is normal.

The critical values for G for $P = 0.05$ are given in Table A.6. If the calculated value of G exceeds the critical value, the suspect value is rejected.

EXAMPLE 3.7.3

Apply Grubbs' test to the data from the previous example.

The seven values have $\bar{x} = 0.4026$, $s = 0.01121$

$$G = |0.380 - 0.4026|/0.01121 = 2.016$$

The critical value ($P = 0.05$) is 2.02. The suspect value 0.380 is just not rejected, unlike Dixon's test: such contradictory results are not unusual in outlier tests.

It is important to appreciate that for a significance level of 5% there is still a chance of 5%, or 1 in 20, of incorrectly rejecting the suspect value. This may have a considerable effect on the estimation of the precision of an experiment. For example, for all seven values of the nitrite concentration given above, the standard deviation is $0.0112\,\text{mg}\,\text{l}^{-1}$ but when the suspect value is rejected the standard deviation becomes $0.0056\,\text{mg}\,\text{l}^{-1}$, i.e. the precision appears to have improved by a factor of 2.

The example above illustrates the importance of caution in rejecting outliers. When measurements are repeated only a few times (which is common in analytical work), rejection of one value makes a great difference to the mean and standard deviation. In particular, the practice of making three measurements and rejecting the one which differs most from the other two should be avoided. It can be shown that a more reliable estimate of the mean is obtained, on average, by using the middle one of the three values rather than the mean of the two unrejected values.

If a set of data contains two or more suspect results, other complications arise in deciding whether rejection is justified. Figure 3.1 illustrates in the form of dot-plots two examples of such difficulties. In Figure 3.1 (a) there are two results (2.9, 3.1), both of which are suspiciously high compared with the mean of the data, yet if Q were calculated uncritically using equation (3.8) we would obtain:

$$Q = (3.1 - 2.9)/(3.1 - 2.0) = 0.18$$

a value which is not significant ($P = 0.05$). Clearly the possible outlier 3.1 has been **masked** by the other possible outlier, 2.9, giving a low value of Q. A different situation is shown in Figure 3.1(b), where the two suspect values

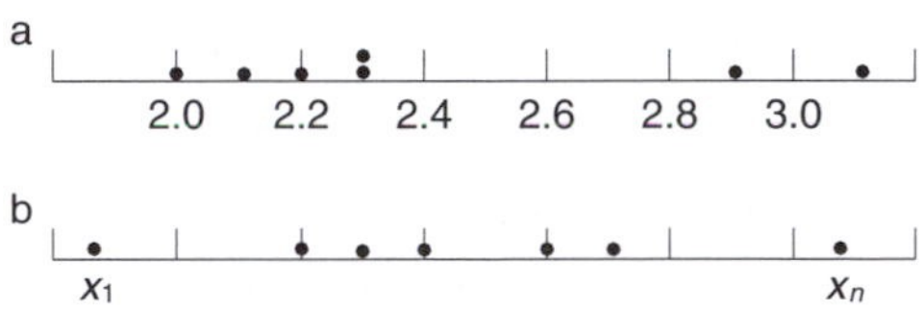

Figure 3.1 Dot-plots illustrating the problem of handling outliers: (a) when there are two suspect results at the high end of the sample data; and (b) when there are two suspect results, one at each extreme of the data.

are at opposite ends of the data set. This results in a large value for the range. As a result Q is small and so not significant. Extensions of Grubbs' test give tests for pairs of outliers. Further details for dealing with multiple outliers can be found from the references given at the end of this chapter.

The outlier tests described above assume that the sample comes from a normal population. It is important to realize that a result that seems to be an outlier on the assumption of a normal population distribution may well not be an outlier if the sample actually comes from (for example) a log-normal distribution (Section 2.3). Therefore outlier tests should not be used if there is a suspicion that the population may not have a normal distribution. This difficulty, along with the extra complications arising in cases of multiple outliers, explains the increasing use of the non-parametric and robust statistical methods described in Chapter 6. Such methods are either insensitive to extreme values, or at least give them less weight in calculations, so the problem of whether or not to reject outliers is avoided.

3.8 Analysis of variance

In Section 3.3 a method was described for comparing two means to test whether they differ significantly. In analytical work there are often more than two means to be compared. Some possible situations are: comparing the mean concentration of protein in solution for samples stored under different conditions; comparing the mean results obtained for the concentration of an analyte by several different methods; and comparing the mean titration results obtained by several different experimentalists using the same apparatus. In all these examples there are two possible sources of variation. The first, which is always present, is due to the random error in measurement. This was discussed in detail in the previous chapter: it is this error which causes a different result to be obtained each time a measurement is repeated under the same conditions. The second possible source of variation is due to what is known as a **controlled** or **fixed-effect factor**. For the examples above, the controlled factors are respectively: the conditions under which the solution was stored, the method of analysis used, and the experimentalist carrying out the titration. Analysis of variance (frequently abbreviated to **ANOVA**) is an extremely powerful statistical technique which can be used to separate and estimate the different causes of variation. For the particular examples above, it can be used to separate any variation which is caused by changing the controlled factor from the variation due to random error. It can thus test whether altering the controlled factor leads to a significant difference between the mean values obtained.

ANOVA can also be used in situations where there is more than one source of random variation. Consider, for example, the purity testing of a barrelful of sodium chloride. Samples are taken from different parts of the barrel chosen at random and replicate analyses performed on these samples. In addition to the random error in the measurement of the purity, there may also be variation in the purity of the samples from different parts of the barrel. Since the samples were chosen at random, this variation will be random and is thus sometimes known as a **random-effect factor**. Again, ANOVA can be used to separate and estimate the sources of variation. Both types of statistical analysis described above, i.e. where there is one factor, either controlled or random, in addition to the random error in measurement, are known as one-way ANOVA. The arithmetical procedures are similar in the fixed- and random-effect factor cases: examples of the former are given in this chapter and of the latter in the next chapter, where sampling is considered in more detail. More complex situations in which there are two or more factors, possibly interacting with each other, are considered in Chapter 7.

3.9 Comparison of several means

Table 3.2 shows the results obtained in an investigation into the stability of a fluorescent reagent stored under different conditions. The values given are the fluorescence signals (in arbitrary units) from dilute solutions of equal concentration. Three replicate measurements were made on each sample. The table shows that the mean values for the four samples are different. However, we know that because of random error, even if the true value which we are trying to measure is unchanged, the sample mean may vary from one sample to the next. ANOVA tests whether the difference between the sample means is too great to be explained by the random error.

Figure 3.2 shows a dot-plot comparing the results obtained in the different conditions. This suggests that there is little difference between conditions A and B but that conditions C and D differ both from A and B and from each other.

The problem can be generalized to consider h samples each with n members as in Table 3.3 where x_{ij} is the jth measurement of the ith sample. The means of the samples are $\bar{x}_1, \bar{x}_2, \ldots, \bar{x}_h$ and the mean of all the values grouped together is $\bar{x}$. The null hypothesis adopted is that all the samples are drawn from a population with mean μ and variance σ_0^2. On the basis of this hypothesis σ_0^2 can be estimated in two ways, one involving the variation *within* the samples and the other the variation *between* the samples.

Table 3.2 Fluorescence from solutions stored under different conditions

Conditions	*Replicate measurements*	*Mean*
A Freshly prepared	102, 100, 101	101
B Stored for 1 hr in the dark	101, 101, 104	102
C Stored for 1 hr in subdued light	97, 95, 99	97
D Stored for 1 hr in bright light	90, 92, 94	92
	Overall mean	98

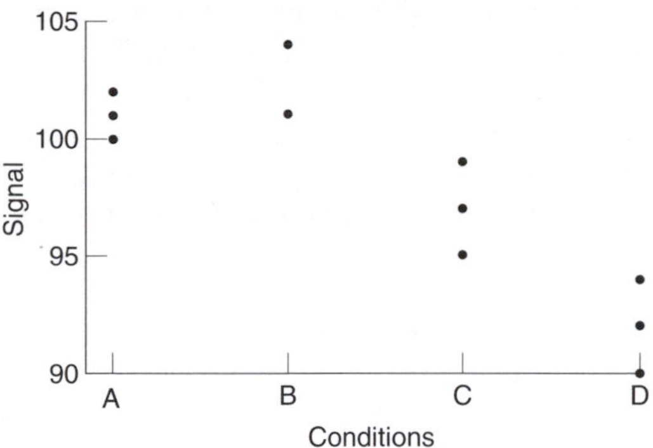

Figure 3.2 Dot-plot of results in Table 3.2.

1. *Within-sample variation*
   For each sample a variance can be calculated by using the formula:

$$\sum(x_i - \bar{x})^2/(n-1) \text{[see equation (2.2)]}$$

Using the values in Table 3.2 we have:

$$\text{Variance of sample A} = \frac{(102-101)^2 + (100-101)^2 + (101-101)^2}{(3-1)} = 1$$

$$\text{Variance of sample B} = \frac{(101-102)^2 + (101-102)^2 + (104-102)^2}{(3-1)} = 3$$

Similarly it can be shown that samples C and D both have variances of 4.
   Averaging these values gives:

$$\text{Within-sample estimate of } \sigma_0^2 = (1+3+4+4)/4 = 3$$

This estimate has 8 degrees of freedom: each sample estimate has 2 degrees of freedom and there are 4 samples. Note that this estimate does not depend on the means of the samples: for example, if all the measurements for sample A were increased by, say, 4, this estimate of σ_0^2 would be unaltered.
   The general formula for the within-sample estimate of σ_0^2 is:

$$\text{Within-sample estimate of } \sigma_0^2 = \sum_i \sum_j (x_{ij} - \bar{x}_i)^2/h(n-1) \qquad (3.10)$$

Table 3.3 Generalization of Table 3.2

			Mean
Sample 1	x_{11}	$x_{12} \ldots x_{1j} \ldots x_{1n}$	$\bar{x}_1$
Sample 2	x_{21}	$x_{22} \ldots x_{2j} \ldots x_{2n}$	$\bar{x}_2$
	$\vdots$	$\vdots \quad \vdots \quad \vdots$	$\vdots$
Sample i	x_{i1}	$x_{i2} \ldots x_{ij} \ldots x_{in}$	$\bar{x}_i$
	$\vdots$	$\vdots \quad \vdots \quad \vdots$	$\vdots$
Sample h	x_{h1}	$x_{h2} \ldots x_{hj} \ldots x_{hn}$	$\bar{x}_h$
		Overall mean =	$\bar{x}$

The summation over j and division by $(n-1)$ gives the variance of each sample; the summation over i and division by h averages these sample variances. The expression in equation (3.10) is known as a **mean square** (MS) since it involves a sum of squared terms (SS) divided by the number of degrees of freedom. Since in this case the number of degrees of freedom is 8 and the mean square is 3, the sum of the squared terms is $3 \times 8 = 24$.

2. *Between-sample variation*
   If the samples are all drawn from a population which has variance σ_0^2, then their means come from a population with variance σ_0^2/n (cf. the sampling distribution of the mean, Section 2.5). Thus, if the null hypothesis is true, the variance of the means of the samples gives an estimate of σ_0^2/n. From Table 3.2:

   Sample mean variance

   $$= \frac{(101-98)^2 + (102-98)^2 + (97-98)^2 + (92-98)^2}{(4-1)}$$

   $$= 62/3$$

   So the estimate of σ_0^2 is $(\frac{62}{3}) \times 3 = 62$. This estimate has 3 degrees of freedom since it is calculated from four sample means. Note that this estimate of σ_0^2 does not depend on the variability within each sample, since it is calculated from the sample means.

   However if, for example, the mean of sample D was changed, then this estimate of σ_0^2 would also be changed.

   In general we have:

   $$\text{Between-sample estimate of } \sigma_0^2 = n\sum_i (\bar{x}_i - \bar{x})^2/(h-1) \qquad (3.11)$$

   which is again a 'mean square' involving a sum of squared terms divided by the number of degrees of freedom. In this case the number of degrees of freedom is 3 and the mean square is 62, so the sum of the squared terms is $3 \times 62 = 186$.

Summarizing our calculations so far:

Within-sample mean square = 3 with 8 d.f.

Between-sample mean square = 62 with 3 d.f.

If the null hypothesis is correct, then these two estimates of σ_0^2 should not differ significantly. If it is incorrect, the between-sample estimate of σ_0^2 will be greater than the within-sample estimate because of between-sample variation. To test whether it is significantly greater, a one-sided F-test is used (see Section 3.6):

$$F = 62/3 = 20.7$$

(Remember each mean *square* is used so no further squaring is necessary.) From Table A.3 the critical value of F is 4.066 ($P = 0.05$). Since the calculated value of F is greater than this the null hypothesis is rejected: the sample means do differ significantly.

A significant result in one-way ANOVA can arise for several different reasons: for example, one mean may differ from all the others, all the means may differ from each other, the means may fall into two distinct groups, etc. A simple way of deciding the reason for a significant result is to arrange the means in increasing order and compare the difference between adjacent values with a quantity called the **least significant difference**. This is given by:

$$s\sqrt{\left(\frac{2}{n}\right)} \times t_{h(n-1)}$$

where s is the within-sample estimate of σ_0 and $h(n-1)$ is the number of degrees of freedom of this estimate. For the example above, the sample means arranged in increasing order of size are:

$$\bar{x}_D = 92 \qquad \bar{x}_C = 97 \qquad \bar{x}_A = 101 \qquad \bar{x}_B = 102$$

and the least significant difference is $\sqrt{3} \times \sqrt{2/3} \times 2.306$ ($P = 0.05$), giving 3.26. Comparing this value with the differences between the means suggests that conditions D and C gives results which differ significantly from each other and from the results obtained in conditions A and B. However the results obtained in conditions A and B do not differ significantly from each other. This confirms what was suggested by the dot plot in Figure 3.2 and suggests that it is exposure to light which affects the fluorescence.

The least significant difference method described above is not entirely rigorous: it can be shown that it leads to rather too many significant differences. However it is a simple follow-up test when ANOVA has indicated that there is a significant difference between the means. Descriptions of other more rigorous tests are given in the references at the end of this chapter.

3.10 The arithmetic of ANOVA calculations

In the preceding ANOVA calculation σ_0^2 was estimated in two different ways. If the null hypothesis were true, σ_0^2 could also be estimated in a third way by treating the data as one large sample. This would involve summing the squares of the deviations from the overall mean:

$$\sum_i \sum_j (x_{ij} - \bar{x})^2 = 4^2 + 2^2 + 3^2 + 3^2 + 3^2 + 6^2 + 1^2 + 3^2 + 1^2 + 8^2 + 6^2 + 4^2$$
$$= 210$$

and dividing by the number of degrees of freedom, $12 - 1 = 11$.

This method of estimating σ_0^2 is not used in the analysis because the estimate depends both on the within- and between-sample variation. However, there is an exact algebraic relationship between this total variation and the sources of variation which contribute to it. This, especially in more complicated ANOVA calculations, leads to a simplification of the arithmetic

Table 3.4 Summary of sums of squares and degrees of freedom

Source of variation	*Sum of squares*	*Degrees of freedom*
Between-sample	$n\sum_i(\bar{x}_i - \bar{x})^2 = 186$	$h - 1 = 3$
Within-sample	$\sum_i\sum_j(x_{ij} - \bar{x}_i)^2 = 24$	$h(n - 1) = 8$
Total	$\sum_i\sum_j(x_{ij} - \bar{x})^2 = 210$	$hn - 1 = 11$

involved. The relationship between the sources of variation is illustrated by Table 3.4, which summarizes the sums of squares and degrees of freedom. It will be seen that the values for the total variation given in the last row of the table are the sums of the values in the first two rows for both the sum of squares and the degrees of freedom. This additive property holds for all the ANOVA calculations described in this book.

Just as in the calculation of variance, there are formulae which simplify the calculation of the individual sums of squares. These formulae are summarized below:

> One-way ANOVA tests for a significant difference between means when there are more than two samples involved. The formulae used are:
>
> | *Source of variation* | *Sum of squares* | *Degrees of freedom* |
> |---|---|---|
> | Between-samples | $\sum_i T_i^2/n - T^2/N$ | $h - 1$ |
> | Within-samples | by subtraction | by subtraction |
> | Total | $\sum_i\sum_j x_{ij}^2 - T^2/N$ | $N - 1$ |
>
> where $N = nh$ = total number of measurements
>
> T_i = sum of the measurements in the ith sample
> T = sum of all the measurements, grand total.
>
> The test statistic is F = between-sample mean square/within-sample mean square and the critical value is $F_{h-1,N-h}$.

These formulae can be illustrated by repeating the ANOVA calculations for the data in Table 3.2. The calculation is given in full below.

EXAMPLE 3.10.1

Test whether the samples in Table 3.2 are drawn from populations with equal means.

The calculation of the mean squares is set out below. All the values in Table 3.2 have had 100 subtracted from them, which simplifies the arithmetic considerably.

Note that this does not affect either the between- or within-sample estimates of variance because the same quantity has been subtracted from every value.

				T_i	T_i^2
A	2	0	1	3	9
B	1	1	4	6	36
C	−3	−5	−1	−9	81
D	−10	−8	−6	−24	576
				$T = -24$	$\sum_i T_i^2 = 702$

$n = 3$, $h = 4$, $N = 12$, $\sum_i \sum_j x_{ij}^2 = 258$

Source of variation	*Sum of squares*	*Degrees of freedom*	*Mean square*
Between-sample	$702/3 - (-24)^2/12 = 186$	3	$186/3 = 62$
Within-sample	by subtraction $= 24$	8	$24/8 = 3$
Total	$258 - (-24)^2/12 = 210$	11	
	$F = 62/3 = 20.7$		

The critical value $F_{3,8} = 4.066$ ($P = 0.05$). Since the calculated value is greater than this the null hypothesis is rejected: the sample means differ significantly.

The calculations for one-way ANOVA have been given in detail in order to make the principles behind the method clearer. In practice such calculations are normally made on a computer. Both Minitab and Excel have an option which performs one-way ANOVA and, as an example, the output given by Excel is shown below, using the original values.

Anova: single factor

SUMMARY

Groups	Count	Sum	Average	Variance
A	3	303	101	1
B	3	306	102	3
C	3	291	97	4
D	3	276	92	4

ANOVA

Source of Variation	SS	df	MS	F	*P*-value	F crit
Between Groups	186	3	62	20.66667	0.0004	4.06618
Within Groups	24	8	3			
Total	210	11				

Certain assumptions have been made in performing the ANOVA calculations in this chapter. The first is that the variance of the random error is not affected by the treatment used. This assumption is implicit in the pooling of the within-sample variances to calculate an overall estimate of the error variance. In doing this we are assuming what is known as the **homogeneity of variance**. In the particular example given above, where all the measurements are made in the same way, we would expect homogeneity of variance. Methods of testing for this property are given in the references at the end of this chapter.

A second assumption is that the uncontrolled variation is random. This would not be the case if, for example, there were some uncontrolled factor, such as temperature change, which produced a trend in the results over a period of time. The effect of such uncontrolled factors can be overcome to a large extent by the techniques of randomization and blocking which are discussed in Chapter 7.

It will be seen that an important part of ANOVA is the application of the F-test. Use of this test (see Section 3.6) simply to compare the variances of two samples depends on the samples being drawn from a normal population. Fortunately, however, the F-test as applied in ANOVA is not too sensitive to departures from normality of distribution.

3.11 The chi-squared test

In the significance tests so far described in this chapter the data have taken the form of observations which, apart from any rounding off, have been measured on a continuous scale. In contrast, this section is concerned with *frequency*, i.e. the number of times a given event occurs. For example, Table 2.2 gives the frequencies of the different values obtained for the nitrate ion concentration when 50 measurements were made on a sample. As discussed in Chapter 2, such measurements are usually assumed to be drawn from a population which is normally distributed. The **chi-squared test** could be used to test whether the observed frequencies differ significantly from those which would be expected on this null hypothesis.

To test whether the observed frequencies, O_i, agree with those expected, E_i, according some null hypothesis, the statistic X^2 is calculated:

$$X^2 = \sum_i \frac{(O_i - E_i)^2}{E_i} \qquad (3.12)$$

Since the calculation involved in using this statistic to test for normality is relatively complicated, it will not be described here. (A reference to a worked example is given at the end of the chapter.) The principle of the chi-squared test is more easily understood by means of the following example.

EXAMPLE 3.11.1

The numbers of glassware breakages reported by four laboratory workers over a given period are shown below. Is there any evidence that the workers differ in their reliability?

Numbers of breakages: 24, 17, 11, 9

The null hypothesis is that there is no difference in reliability. Assuming that the workers use the laboratory for an equal length of time, we would expect, from the null hypothesis, the same number of breakages by each worker. Since the total number of breakages is 61, the expected number of breakages per worker is $61/4 = 15.25$. Obviously it is not possible in practice to have a non-integral number of breakages: this number is a mathematical concept. The nearest practicable 'equal' distribution is 15, 15, 15, 16, in some order. The question to be answered is whether the difference between the observed and expected frequencies is so large that the null hypothesis should be rejected. That there should be *some* difference between the two sets of frequencies can be most readily appreciated by considering a sequence of throws of a die: we should, for example, be most surprised if 30 throws of a die yielded exactly equal frequencies for 1, 2, 3, etc. The calculation of X^2 is shown below.

Observed frequency, O	*Expected frequency, E*	$O - E$	$(O - E)^2/E$
24	15.25	8.75	5.020
17	15.25	1.75	0.201
11	15.25	−4.25	1.184
9	15.25	−6.25	2.561
Totals	61	0	$X^2 = 8.966$

Note that the total of the $O - E$ column is always zero, thus providing a useful check on the calculation.

If X^2 exceeds a certain critical value the null hypothesis is rejected. The critical value depends, as in other significance tests, on the significance level of the test and on the number of degrees of freedom. The number of degrees of freedom is, in an example of this type, one less than the number of classes used, i.e. $4 - 1 = 3$ in this case. The critical values of X^2 for $P = 0.05$ are given in Table A.7. For 3 degrees of freedom the critical value is 7.81. Since the calculated value is greater than this the null hypothesis is rejected at the 5% significance level: there is evidence that the workers *do* differ in their reliability.

The calculation of X^2 suggests that a significant result is obtained because of the high number of breakages reported by the first worker. To study this further, additional chi-squared tests can be performed. One of them tests whether the second, third and fourth workers differ significantly from each other: in this case each expected frequency is $(17 + 11 + 9)/3$. (Note that the t-test cannot be used here as we are dealing with frequencies and not continuous variates). Another tests whether the first worker differs significantly from the other three workers taken as a group. In this case there are two classes: the breakages by the first worker with an expected frequency of 15.25 and the total breakages by the other workers with expected frequency of $15.25 \times 3 = 45.75$. In such cases when there are only two classes and hence one degree of freedom, an adjustment known as **Yates's correction** should be applied. This involves replacing $O - E$ by $|O - E| - 0.5$. For example, if $O - E = -4.5$, $|O - E| = 4.5$ and $|O - E| - 0.5 = 4$. These further tests are given as an exercise at the end of this chapter.

In general the chi-squared test should only be used if the total number of observations is 50 or more and the individual expected frequencies are not less than 5. This is not a rigid rule: a reference is given at the end of this chapter which discusses this point further. Other applications of the chi-squared test are also described in this reference.

3.12 Testing for normality of distribution

As has been emphasized in this chapter, many statistical tests assume that the data used are drawn from a normal population. One method of testing this assumption, using the chi-squared test, was mentioned in the previous section. Unfortunately, this method can only be used if there are 50 or more data points. It is common in experimental work to have only a small set of data. A simple visual way of seeing whether a set of data is consistent with the assumption of normality is to plot a **cumulative frequency curve** on special graph paper known as **normal probability paper**. This method is most easily explained by means of an example.

EXAMPLE 3.12.1

Use normal probability paper to investigate whether the data below could have been drawn from a normal population:

109, 89, 99, 99, 107, 111, 86, 74, 115, 107, 134, 113, 110, 88, 104

Table 3.5 Data for normal probability paper example

Measurement	*Cumulative frequency*	*% Cumulative frequency*
74	1	6.25
86	2	12.50
88	3	18.75
89	4	25.00
99	6	37.50
104	7	43.75
107	9	56.25
109	10	62.50
110	11	68.75
111	12	75.00
113	13	81.25
115	14	87.50
134	15	93.75

Table 3.5 shows the data arranged in order of increasing size. The second column gives the cumulative frequency for each measurement, i.e. the number of measurements less than or equal to that measurement. The third column gives the percentage cumulative frequency. This is calculated by using the formula:

$$\%\ \text{cumulative frequency} = 100 \times \text{cumulative frequency}/(n+1)$$

where n is the total number of measurements. (A divisor of $n+1$ rather than n is used so that the % cumulative frequency of 50% falls at the middle of the data set, in this case at the eighth measurement.) If the data come from a normal population, a graph of percentage cumulative frequency against measurement results in an S-shaped curve, as shown in Figure 3.3.

Normal probability paper has a non-linear scale for the percentage cumulative frequency axis, which will convert this S-shaped curve into a straight line. A graph plotted on such paper is shown in Figure 3.4: the points lie approximately on a straight line, supporting the hypothesis that the data come from a normal distribution.

Minitab will give a normal probability plot directly. There is a choice of three different algorithms for calculating the cumulative frequencies in Table 3.5. The algorithm used above is known as the Herd–Johnson method.

One method of testing for normality is to use a quantity which measures how closely the points on a normal probability plot conform to a straight line. The calculation of this quantity, the product–moment correlation coefficient, r, is described in a later chapter (Section 5.3). A reference to the use of r for testing for normality is given at the end of this chapter. This reference also gives a survey of the different tests for normality. Section 6.12 describes another method, the Kolmogorov–Smirnov method, which, among other applications, may be used to test for normality. A worked example is given in that section.

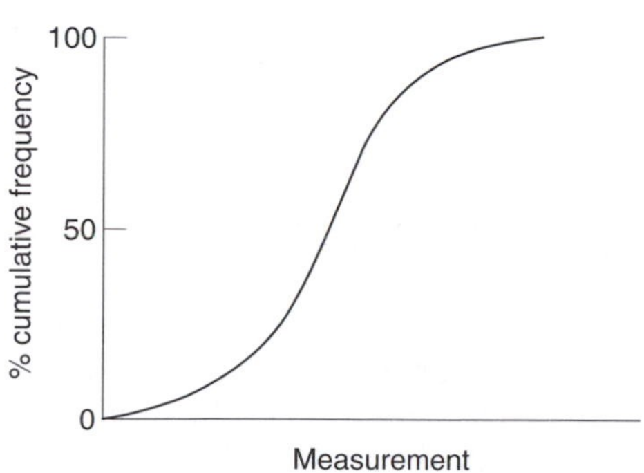

Figure 3.3 The cumulative frequency curve for a normal distribution.

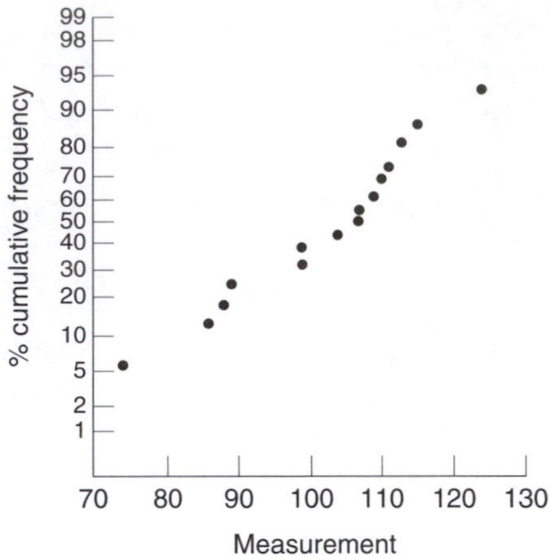

Figure 3.4 Normal probability plot for the example in Section 3.12.

3.13 Conclusions from significance tests

This section looks more closely at the conclusions which may be drawn from a significance test. As was explained in Section 3.2, a significance test at, for example, the $P = 0.05$ level involves a 5% risk that a null hypothesis will be rejected *even though it is true*. This type of error is known as a **Type I error**. The risk of such an error can be reduced by altering the significance level of the test to $P = 0.01$ or even $P = 0.001$. This, however, is not the only possible type of error: it is also possible to retain a null hypothesis even when it is false. This is called a **Type II error**. In order to calculate the probability of this type of error it is necessary to postulate an alternative to the null hypothesis, known as an **alternative hypothesis, H_1**.

Consider the situation where a certain chemical product is meant to contain 3% of phosphorus by weight. It is suspected that this proportion has increased. To test this the composition is analysed by a standard method with a known standard deviation of 0.036%. Suppose four measurements are made and a significance test is performed at the $P = 0.05$ level. A one-sided test is required, as we are interested only in an increase. The null hypothesis is:

$$H_0: \quad \mu = 3.0\%$$

The solid line in Figure 3.5 shows the sampling distribution of the mean if H_0 is true. This sampling distribution has mean 3.0 and standard deviation (i.e. standard error of the mean) $\sigma/\sqrt{n} = 0.036/\sqrt{4}\%$. If the sample mean lies above the indicated critical value, $\bar{x}_c$, the null hypothesis is rejected. Thus the black region, with area 0.05, represents the probability of a Type I error.

Suppose we take the alternative hypothesis:

$$H_1: \quad \mu = 3.05\%.$$

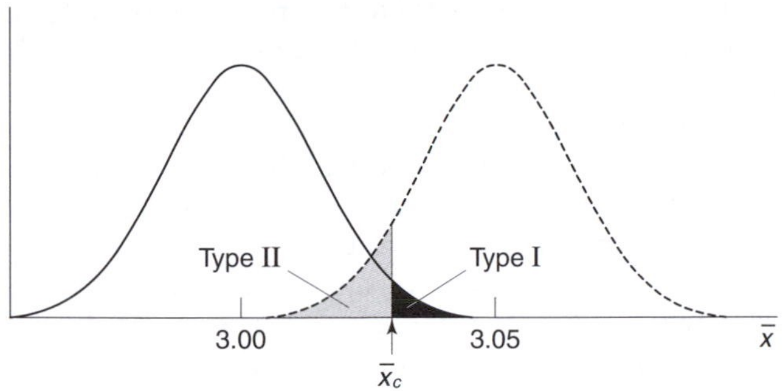

Figure 3.5 Type I and Type II errors.

The broken line in Figure 3.5 shows the sampling distribution of the mean if the alternative hypothesis is true. *Even if this is the case*, the null hypothesis will be retained if the sample mean lies below $\bar{x}_c$. The probability of this Type II error is represented by the shaded area. The diagram shows the interdependence of the two types of error. If, for example, the significance level is changed to $P = 0.01$ in order to reduce a risk of a Type I error, $\bar{x}_c$ will be increased and the risk of a Type II error is also increased. Conversely, a decrease in the risk of a Type II error can only be achieved at the expense of an increase in the probability of a Type I error. The only way in which both errors can be reduced (for a given alternative hypothesis) is by increasing the sample size. The effect of increasing n to 9, for example, is illustrated in Figure 3.6: the resultant decrease in the standard error of the mean produces a decrease in both types of error, for a given value of $\bar{x}_c$.

The probability that a false null hypothesis is rejected is known as the **power** of a test. That is, the power of a test is (1 – the probability of a Type II error). In the example above it is a function of the mean specified in the alternative hypothesis. It also depends on the sample size, the significance level of the test, and whether the test is one- or two-sided. In some circumstances

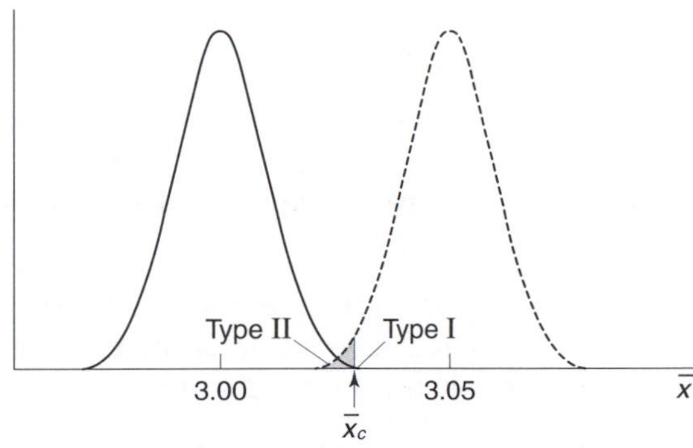

Figure 3.6 Type I and Type II errors for increased sample size.

where two or more tests are available to test the same hypothesis, it may be useful to compare the powers of the tests in order to decide which is most appropriate.

Type I and Type II errors are also relevant when significance tests are applied sequentially. An example of this situation is the application of the t-test to the difference between two means, after first using the F-test to decide whether or not the sample variances can be pooled (see Sections 3.3 and 3.6). Both Type I and Type II errors can arise from the initial F-test, and the occurrence of either type will mean that the stated levels of significance for the subsequent t-test are incorrect, because the incorrect form of the t-test may have been applied.

This example emphasizes the general conclusion that significance tests do not give clear-cut answers: rather they aid the interpretation of experimental data by giving the probabilities that certain conclusions are valid.

Bibliography

Barnett, V. and Lewis, T. 1994. *Outliers in Statistical Data*. 3rd Edn. Wiley, Chichester. (A very comprehensive treatment of the philosophy of outlier rejection and the tests used.)

Box, G. E. P., Hunter, W. G. and Hunter, J. S. 1978. *Statistics for Experimenters*. Wiley, New York. (Gives further details of testing for significant differences between means as a follow-up to ANOVA.)

Campbell, R. C. 1989. *Statistics for Biologists*. 3rd Edn. Cambridge University Press, Cambridge. (Gives tests for homogeneity of variance and normality.)

Crawshaw, J. and Chambers, J. 1997. *A Concise Course in A-Level Statistics*. 3rd Edn. Stanley Thornes, Cheltenham. (Gives examples of the chi-squared test for normality.)

Davies, O. L. and Goldsmith, P. L. 1984. *Statistical Methods in Research and Production*. 4th Edn. Longman, London. (Gives more detail about Type I and Type II errors and other applications of the chi-squared test.)

Filliben, J. J. 1975. *Technometrics* 17: 111. (Describes the use of r in testing for normality, and surveys other tests for normality.)

Kleinbaum, D. G., Kupper, L. L. and Muller, K. E. 1988. *Applied Regression Analysis and Other Multivariable Methods*. 2nd Edn. PWS – Kent Publishing, Boston. (Gives further details of testing for significant differences between means as a follow-up to ANOVA.)

Kowalski, B. R. (ed.). 1977. *Chemometrics: Theory and Application*. American Chemical Society, Washington. (Chapter 11 describes tests for normality and discusses the effect of non-normality on parametric tests.)

Sokat, R. R. and Rohlf, F. J. 1994. *Biometry*. 3rd Edn. Freeman, New York. (Gives details of tests for homogeneity of variance.)

1. Use a normal probability plot to test whether the following set of data could have been drawn from a normal population:

   11.68, 11.12, 8.92, 8.82, 10.31, 11.88, 9.84, 11.69, 9.53, 10.30, 9.17, 10.04, 10.65, 10.91, 10.32, 8.71, 9.83, 9.90, 10.40

2. In order to evaluate a spectrophotometric method for the determination of titanium, the method was applied to alloy samples containing different certified amounts of titanium. The results (% Ti) are shown below.

Sample	*Certified value*	*Mean*	*Standard deviation*
1	0.496	0.482	0.0257
2	0.995	1.009	0.0248
3	1.493	1.505	0.0287
4	1.990	2.002	0.0212

   For each alloy eight replicate determinations were made.

   (Qiu Xing-chu and Zhu Ying-quen. 1983. *Analyst* 108: 641)

   For each alloy, test whether the mean value differs significantly from the certified value.

3. For the data in Example 3.3.3, concerning the concentration of thiol in blood lysate,

   (a) Verify that 2.07 is not an outlier for the 'normal' group.
   (b) Show that the variances of the two groups differ significantly.

4. The following data give the recovery of bromide from spiked samples of vegetable matter, measured by using a gas–liquid chromatographic method. The same amount of bromide was added to each specimen.

Tomato:	777	790	759	790	770	758	764 $\mu g\,g^{-1}$
Cucumber:	782	773	778	765	789	797	782 $\mu g\,g^{-1}$

   (Roughan, J. A., Roughan, P. A. and Wilkins, J. P. G. 1983. *Analyst* 108: 742)

   (a) Test whether the recoveries from the two vegetables have variances which differ significantly.
   (b) Test whether the mean recovery rates differ significantly.

5. The following results show the percentage of the total available interstitial water recovered by centrifuging samples taken at different depths in sandstone.

Depth of sample (m)	*Water recovered (%)*					
7	33.3	33.3	35.7	38.1	31.0	33.3
8	43.6	45.2	47.7	45.4	43.8	46.5
16	73.2	68.7	73.6	70.9	72.5	74.5
23	72.5	70.4	65.2	66.7	77.6	69.8

   (Wheatstone, K. G. and Getsthorpe, D. 1982. *Analyst* 107: 731)

Show that the percentage of water recovered differs significantly at different depths. Use the least significant difference method described in Section 3.9 to find the causes of this significant result.

6. The following table gives the concentration of norepinephrine (μmol per gram creatinine) in the urine of healthy volunteers in their early twenties.

Male	0.48	0.36	0.20	0.55	0.45	0.46	0.47	0.23
Female	0.35	0.37	0.27	0.29				

(Yamaguchi, M., Ishida, J. and Yoshimura, M. 1998. *Analyst* 123: 307)

Is there any evidence that concentration of norepinephrine differs between the sexes?

7. In reading a burette to 0.01 ml the final figure has to be estimated. The following frequency table gives the final figures of 50 such readings. Carry out an appropriate significance test to determine whether some digits are preferred to others.

Digit	0	1	2	3	4	5	6	7	8	9
Frequency	1	6	4	5	3	11	2	8	3	7

8. The following table gives further results from the paper cited in Example 3.3.1 (Sahuquillo, A., Rubio, R. and Rauret, G. 1999. *Analyst* 124**)**, in which the results of the determination of chromium in organic materials were compared for two different methods.

Pine needles:	Method 1	mean = 2.15	s.d. = 0.26
	Method 2	mean = 2.45	s.d. = 0.14
Beech leaves:	Method 1	mean = 5.12	s.d. = 0.80
	Method 2	mean = 7.27	s.d. = 0.44
Aquatic plant:	Method 1	mean = 23.08	s.d. = 2.63
	Method 2	mean = 32.01	s.d. = 4.66

In each case the mean is the average of five values.

For each material test whether the mean results obtained by the two methods differ significantly.

9. The data given in the example in Section 3.11 for the number of breakages by four different workers are reproduced below:

24, 17, 11, 9

Test whether:

(a) The number of breakages by the first worker differs significantly from those of the other three workers.

(b) The second, third and fourth workers differ significantly from each other in carefulness.

10. A new flow injection analysis enzymatic procedure for determining hydrogen peroxide in water was compared with a conventional method involving redox titration with potassium permanganate by applying both methods to samples of peroxide for pharmaceutical use. The table

below gives the amount of hydrogen peroxide found, in mg ml^{-1}. Each value is the mean of four replicate measurements.

Sample no.	*Enzymatic method*	*Permanganate method*
1	31.1	32.6
2	29.6	31.0
3	31.0	30.3

(da Cruz Vieira, I. and Fatibello-Filho, O. 1998. *Analyst* 123: 1809)

Test whether the results obtained by the two different methods differ significantly.

11. Six analysts each made six determinations of the paracetamol content of the same batch of tablets. The results are shown below:

Analyst	*Paracetamol content (% m/m)*					
A	84.32	84.51	84.63	84.61	84.64	84.51
B	84.24	84.25	84.41	84.13	84.00	84.30
C	84.29	84.40	84.68	84.28	84.40	84.36
D	84.14	84.22	84.02	84.48	84.27	84.33
E	84.50	83.88	84.49	83.91	84.11	84.06
F	84.70	84.17	84.11	84.36	84.61	83.81

(Trafford, A. D., Jee, R. D., Moffat, A. C. and Graham, P. 1999. *Analyst* 124: 163)

Test whether there is any significant difference between the means obtained by the six analysts.

12. The following figures refer to the concentration of albumin, in g l^{-1}, in the blood sera of 16 healthy adults:

37, 39, 37, 42, 39, 45, 42, 39, 44, 40, 39, 45, 47, 47, 43, 41

(Foote, J. W. and Delves, H. T. 1983. *Analyst* 108: 492)

The first eight figures are for men and the second eight for women. Test whether the mean concentrations for men and women differ significantly.

13. A new flame atomic-absorption spectroscopic method of determining antimony in the atmosphere was compared with the recommended calorimetric method. For samples from an urban atmosphere the following results were obtained:

	Antimony found (mg m^{-3})	
Sample no.	*New method*	*Standard method*
1	22.2	25.0
2	19.2	19.5
3	15.7	16.6
4	20.4	21.3
5	19.6	20.7
6	15.7	16.8

(Castillo, J. R., Lanaja, J., Marinez, M. C. and Aznárez, J. 1982. *Analyst* 107: 1488)

Do the results obtained by the two methods differ significantly?

14. For the situation described in Section 3.13 ($H_0: \mu = 3.0\%$, $H_1: \mu = 3.05\%$, $\sigma = 0.036\%$) calculate the minimum size of sample required to make the probability of a Type I error and the probability of a Type II both equal to 0.01 at most.

CHAPTER FOUR

The quality of analytical measurements

4.1 Introduction

As we saw in Chapter 1, analytical chemistry is an applied measurement science in which quantitative studies predominate, and therefore one in which estimates of the inevitable errors are essential. In almost all applications of analysis the results obtained are supplied to a customer or user, and it is necessary that these users are satisfied as far as possible with the **quality** – the fitness for purpose – of the measurements. This has many important implications for analytical practice. Firstly, any assessment of the measurement errors must take into account the whole analytical process – including the sampling steps, which often contribute to the overall error very significantly. Secondly, the performance of the analyses undertaken in each laboratory must be checked internally on a regular basis, usually by applying them to standard or reference materials. Thirdly, in many application areas the results from different laboratories must be compared with each other, so that the users can be satisfied that the performance of the laboratories meets statutory, regulatory and other requirements. Finally, the analytical results must be supplied with a realistic estimate of their uncertainty, i.e. the range within which the true value of the quantity being measured lies. These are the major topics discussed in this chapter. The statistical methods used in such areas are often very simple in principle, most of them being based on techniques described in Chapters 2 and 3. But their more frequent and improved application has been one of the major developments in analytical sciences in recent years, with a correspondingly large improvement in the quality and acceptability of many analytical results. Moreover some of the methods discussed have broader applications. For example, the principles used to monitor the performance of a single analysis in a single laboratory over a period of time can also be applied to the monitoring of an industrial process.

4.2 Sampling

In most analyses we rely on chemical samples to give us information about a whole object. So unless the sampling stages of an analysis are considered carefully, the statistical methods discussed in this book may be invalidated, as the samples studied may not be properly representative of the whole object under study. For example, it is not possible to analyse all the water in a stream for a toxic pollutant, and it is not possible to analyse all the milk in a tanker lorry to see if it contains a prohibited steroid hormone. In other instances a small sample has to be used because the analytical method is destructive, and we wish to preserve the remainder of the material. So in each case the sample studied must be taken in a way that ensures as far as possible that it is truly representative of the whole object.

To illustrate some aspects of sampling let us consider the situation in which we have a large batch of tablets and wish to obtain an estimate for the mean weight of a tablet. Rather than weigh all the tablets, we take a few of them (say ten) and weigh each one. In this example the batch of tablets forms the population and the ten weighed tablets form a sample from this population (see Section 2.2). If the sample is to be used to deduce the properties of the population, it must be what is known statistically as a **random sample**, i.e. a sample taken in such a way that all the members of the population have an equal chance of inclusion. Only then will equations such as (2.9), which gives the confidence limits of the mean, be valid. It must be appreciated that the term 'random' has, in the statistical sense, a different meaning from 'haphazard'. Although in practice an analyst might spread the tablets on his desk and attempt to pick a sample of ten in a haphazard fashion, such a method could conceal an unconscious bias. The best way to obtain a random sample is by the use of a random number table. Each member of the population is allocated a number in such a way that all the numbers have an equal number of digits e.g. 001, 002, 003, etc. Random numbers are then read off from a random number table (see Table A.8), starting at an arbitrary point to give, for example, 964, 173, etc., and the corresponding members of the population form the sample. An alternative (and much simpler) method which is sometimes used is to select the population members at regular intervals, for example to take every hundredth tablet off a production line. This method is not entirely satisfactory since there might be a coinciding periodicity in the weight of the tablets: the importance of the randomness of the sample is evident. Again, if the last few tablets were taken and there had been a gradual decrease in weight during the production of the batch, then this sample would give an entirely erroneous value for the mean weight of the whole batch.

In the example above the population is made up of obvious discrete members that are nominally the same, viz. the tablets. Sampling from materials for which this is not true, such as rocks, powders, gases and liquids is called **bulk sampling**. If a bulk material were perfectly homogeneous then only a small portion or **test increment** would be needed to determine the properties of the bulk. In practice bulk materials are non-homogeneous for a variety of reasons. Materials such as ores and sediments consist of macroscopic particles with different compositions and these may not be uniformly distributed in the bulk. Fluids may be non-homogeneous on a molecular scale owing to concentration gradients. Such inhomogeneity can only be detected by taking a sample of test increments from different parts of the bulk. If possible this should be done randomly by considering the bulk as a collection of cells of equal size and selecting a sample of cells by using random numbers as described above.

From the random sample, the mean, $\bar{x}$ and variance, s^2, can be calculated. There are two contributions to s^2: the **sampling variance**, σ_1^2, due to differences between the members of the sample, e.g. the tablets having different weights, and the **measurement variance**, σ_0^2, e.g. random errors in weighing each tablet. The next section describes how these two contributions can be separated and estimated by using ANOVA. For bulk materials the sampling variance is dependent on the size of the test increment relative to the scale of the inhomogeneities. As the test increment size increases, the inhomogeneities tend to be averaged out and so the sampling variance decreases.

4.3 Separation and estimation of variances using ANOVA

Section 3.8 described the use of one-way ANOVA to test for differences between means when there was a possible variation due to a fixed-effect factor. In this section we consider the situation where there is a random-effect factor, viz. sampling variation. One-way ANOVA is then used to separate and estimate the different sources of variation, rather than to test whether several sample means differ significantly. Table 4.1 shows the results of the purity testing of a barrelful of sodium chloride. Five sample increments, A–E, were taken from different parts of the barrel chosen at random, and four replicate analyses were performed on each sample. As explained above, there are two possible sources of variation: that due to the random error in the measurement of purity, given by the measurement variance, σ_0^2, and that due to real variations in the sodium chloride purity at different points in the barrel, given by the sampling variance, σ_1^2. Since the within-sample mean square does not depend on the sample mean (see

Table 4.1 Purity testing of sodium chloride

Sample	*Purity (%)*	*Mean*
A	98.8, 98.7, 98.9, 98.8	98.8
B	99.3, 98.7, 98.8, 99.2	99.0
C	98.3, 98.5, 98.8, 98.8	98.6
D	98.0, 97.7, 97.4, 97.3	97.6
E	99.3, 99.4, 99.9, 99.4	99.5

Section 3.9) it can be used to give an estimate of σ_0^2. The between-sample mean square *cannot* be used to estimate σ_1^2 directly, because the variation between sample means is caused both by the random error in measurement and by possible variation in the purity. It can be shown that the between-sample mean square gives an estimate of $\sigma_0^2 + n\sigma_1^2$ (where n is the number of replicate measurements). However, before an estimate of σ_1^2 is made, a test should be carried out to see whether it differs significantly from 0. This is done by comparing the within- and between-sample mean squares: if they do not differ significantly then $\sigma_1^2 = 0$ and both mean squares estimate σ_0^2.

The one-way ANOVA output from Excel for this example is shown below. The results show that the between-sample mean square is greater than the within-sample mean square, and the result of the F-test shows that this difference is highly significant, i.e. that σ_1^2 does differ significantly from 0. The within-sample mean square gives 0.0653 as an estimate of σ_0^2, so we can estimate σ_1^2 using:

$$\begin{aligned} \sigma_1^2 &= (\text{between-sample mean square} - \text{within-sample mean square})/n \\ &= (1.96 - 0.0653)/4 \\ &= 0.47 \end{aligned}$$

Sample A	Sample B	Sample C	Sample D	Sample E
98.8	99.3	98.3	98	99.3
98.7	98.7	98.5	97.7	99.4
98.9	98.8	98.8	97.4	99.9
98.8	99.2	98.8	97.3	99.4

One-way Anova

SUMMARY

Groups	*Count*	*Sum*	*Average*	*Variance*
Sample A	4	395.2	98.8	0.006667
Sample B	4	396	99	0.086667
Sample C	4	394.4	98.6	0.06
Sample D	4	390.4	97.6	0.1
Sample E	4	398	99.5	0.073333

Source of variation	*SS*	*df*	*MS*	*F*	*P-value*	*F crit*
Between-sample	7.84	4	1.96	30	5.34E-07	3.056
Within-sample	0.98	15	0.0653			
Total	8.82	19				

If one analysis is made on each of the h test increments (example above, Section 4.3) then the confidence limits of the mean are given by equation (2.9):

$$\mu = \bar{x} \pm t_{n-1} s/\sqrt{n} \qquad (4.1)$$

where $\bar{x}$ is the mean of the n measurements and s^2 is the variance of the measurements; s^2 is an estimate of the total variance, σ^2, which is the sum of the measurement and sampling variances, i.e. $\sigma_0^2 + \sigma_1^2$ (see Section 2.11), and σ^2/h (which is estimated by s^2/h) is the variance of the mean, $\bar{x}$. If the value for each test increment is the mean of n replicate measurements, then the variance of the mean is $(\sigma_0^2/n + \sigma_1^2)/h = \sigma_0^2/nh + \sigma_1^2/h$. Obviously, for maximum precision, we require the variance of the mean to be as small as possible. The term due to the measurement variance can be reduced either by using a more precise method of analysis or by increasing n, the number of replicate measurements. However, there is no point in striving to make the measurement variance much less than say a tenth of the sampling variance, as any further reduction will not greatly improve the total variance (since it is the sum of the two variances). Rather it is preferable to take a larger number of test increments, since the confidence interval decreases with increasing h. If a preliminary sample has been used to estimate s, then the sample size required to achieve a given size of confidence interval can be calculated approximately (see Chapter 2, Exercise 4).

A possible sampling strategy with bulk material is to take h test increments and blend them before making n replicate measurements. The variance of the mean of these replicate measurements is $\sigma_0^2/n + \sigma_1^2/h$. This total variance should be compared with that when each sample increment is analysed n times and the increment means are averaged, the variance then being $\sigma_0^2/nh + \sigma_1^2/h$ (see above). Obviously the latter variance is the smaller, resulting in greater precision of the mean, but more measurements (nh against h) are required. Knowledge of the values of σ_0^2 and σ_1^2 from previous experience, and the costs of sampling and analysis, can be used to calculate the cost of relative sampling strategies. In general the most economical scheme to give the required degree of precision will be used.

For bulk materials the sampling variance depends on the size of the test increment relative to the scale of the inhomogeneities and decreases with increasing sample increment size. In some experiments it may be necessary to set an upper limit on the sampling variance so that changes in the mean can be detected. Preliminary measurements can be made to decide the minimum test increment size required to give an acceptable level of sampling variance.

4.5 Quality control methods – Introduction

If a laboratory is to produce analytical results of a quality that is acceptable to its clients, and allow it to perform well in proficiency tests or collaborative trials (see below), it is obviously essential that the results obtained in that laboratory should show excellent consistency from day to day. Checking for such consistency is complicated by the occurrence of random errors, so several statistical techniques have been developed to show whether or not time dependent trends are occurring in the results, alongside these inevitable random errors. These are referred to as **quality control** methods.

Suppose that a laboratory uses a chromatographic method for determining the level of a pesticide in fruits. The results may be used to determine whether a large batch of fruit is acceptable or not, and their quality is thus of great importance. The performance of the method will be checked at regular intervals by applying it, with a small number of replicate analyses, to a standard reference material (SRM), the pesticide level in which is certified by a regulatory authority. Alternatively an internal quality control (IQC) standard of known composition and high stability can be used. The SRM or IQC standard will probably be inserted at random into the sequence of materials analysed by the laboratory, so that the IQC materials are not separately identified to the laboratory staff and are studied using exactly the same procedures as those used for the routine samples. The known concentration of the pesticide in the SRM/IQC materials is the target value for the analysis, μ_0. The laboratory needs to be able to stop and examine the analytical method if it seems to be giving erroneous results. On the other hand resources, time and materials will be wasted if the sequence of analyses is halted unnecessarily, so the quality control methods should allow its continued use as long as it is working satisfactorily. If the values for the IQR samples do not show significant time-dependent trends, and if the random errors in the measurements are not too large, the analytical process is under control.

Quality control methods are also very widely used to monitor industrial processes. Again it is important to stop a process if its output falls outside certain limits, but it is equally important not to stop the process if it is working well. For example, the weights of pharmaceutical tablets coming off a production line can be monitored by taking small samples (see above) of tablets from time to time. The tablet weights are bound to fluctuate around the target value μ_0 because of random errors, but if these random errors are not too large, and are not accompanied by time dependent trends, the process is under control.

4.6 Shewhart charts for mean values

In Chapter 2 we showed how the mean, $\bar{x}$, of a sample of measurements could be used to provide an estimate of the population mean, μ, and how the sample standard deviation, s, provided an estimate of the population standard deviation, σ. For a small sample size, n, the confidence limits of the mean are normally given by equation (2.9), with the t value chosen according to the number of degrees of freedom $(n - 1)$ and the confidence level required. The same principles can be applied to quality control work, but with one important difference. Over a long period, the *population* standard deviation, σ, of the pesticide level in the fruit (or, in the second example, of the tablet weights), will become known from experience. In quality control work, σ is given the title **process capability**. Equation (2.9) can be replaced by equation (2.8) with the estimate s replaced by the known σ. In practice $z = 1.96$ is often rounded to 2 for 95% confidence limits and $z = 2.97$ is rounded to 3 for 99.7% confidence limits.

$$\text{For 95\% confidence limits:} \quad \mu = \bar{x} \pm \frac{2\sigma}{\sqrt{n}} \tag{4.2}$$

$$\text{For 99.7\% confidence limits:} \quad \mu = \bar{x} \pm \frac{3\sigma}{\sqrt{n}} \tag{4.3}$$

These equations are used in the construction of the most common type of control chart, a **Shewhart chart** (Figure 4.1). The vertical axis of a Shewhart chart displays the **process mean**, $\bar{x}$, of the measured values, e.g. of the pesticide concentration in the fruit, and the horizontal axis is a time axis, so that the variation of these $\bar{x}$ values with time can be plotted. The **target value**, μ_0, is marked by a horizontal line. The chart also includes two further pairs

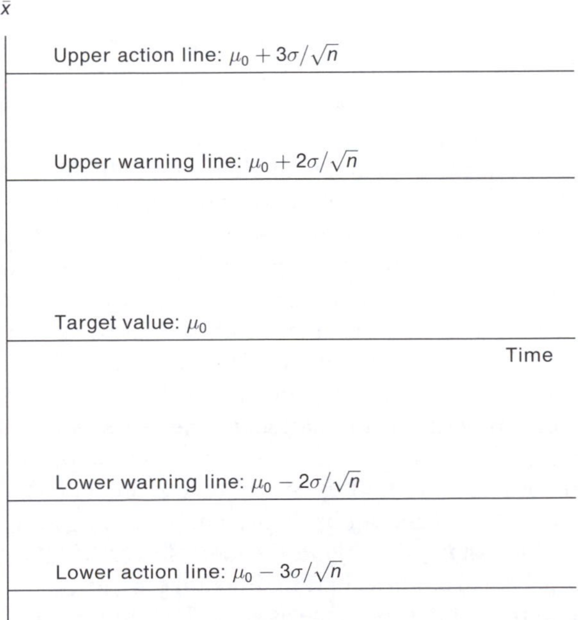

Figure 4.1 Shewhart chart for mean values.

of horizontal lines. The lines at $\mu_0 \pm 2\sigma/\sqrt{n}$ are called the **warning** lines, and those at $\mu_0 \pm 3\sigma/\sqrt{n}$ are called the **action lines**. The purpose of these lines is indicated by their names. Suppose a measured $\bar{x}$ value falls outside the action lines. The probability of such an occurrence when the process is in control is known to be only 0.3%, i.e. 0.003, so in practice the process is usually stopped and examined if this occurs. There is a probability of ca. 5% (0.05) of a single point falling outside either warning line (but within the action lines) while the process remains in control. This alone would not cause the process to be stopped, but if *two successive* points fall outside the same warning line, the probability of such an occurrence ($P = [0.025]^2 \times 2 = 0.00125$ in total for both warning lines) is again so low that the process is judged to be out of control. These two criteria – one point outside the action lines, or two successive points outside the same warning line – are the ones most commonly applied in the interpretation of Shewhart charts. Others are often used in addition: for example the probability of eight successive points lying on one specific side of the target value line is clearly low, i.e. $(0.5)^8 = 0.0039$, and such an occurrence again suggests that the process is out of control. Provision can also be made for stopping a process in cases where the plotted $\bar{x}$ values show a trend (e.g. six increasing or decreasing points in succession, even if the points are within the warning lines), or where they seem to oscillate (e.g. 14 successive points, alternating up and down). Users of control charts must establish clearly all the criteria to be used in declaring their process out of control.

4.7 Shewhart charts for ranges

If a Shewhart chart for mean values suggests that a process is out of control, there are two possible explanations. The most obvious is that the process mean has changed: the detection of such changes is the main reason for using control charts in which $\bar{x}$ values are plotted. An alternative explanation is that the process mean has remained unchanged but that the variation in the process has increased, i.e. that the action and warning lines are too close together, giving rise to indications that changes in $\bar{x}$ have occurred when in fact they have not. Errors of the opposite kind are also possible. If the variability of the process has diminished (i.e. improved), then the action and warning lines will be too far apart, perhaps allowing real changes in $\bar{x}$ to go undetected. Therefore we must monitor the variability of the process as well as its mean value. This monitoring also has its own intrinsic value: the variability of a process or an analysis is one measure of its quality, and in the laboratory situation is directly linked to the repeatability (within-laboratory standard deviation) of the method (see Chapter 1).

The variability of a process can be displayed by plotting another Shewhart chart to display the **range**, R (= highest value − lowest value), of each of the samples taken. A typical control chart for the range is shown in Figure 4.2. The general format of the chart is the same as is used in plotting mean values, with a line representing the target value, and also pairs of action and warning lines.

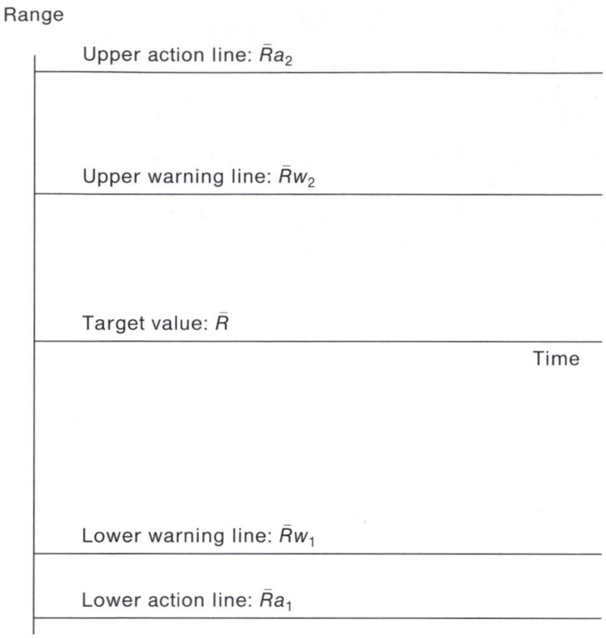

Figure 4.2 Shewhart chart for range.

The most striking difference between the two charts is that these pairs of lines are not symmetrical with respect to the target value for the range, $\bar{R}$. The value of $\bar{R}$ can be calculated using the value of σ, and the positions of the action and warning lines can be derived from $\bar{R}$, using multiplying factors obtained from statistical tables. These factors take values depending on the sample size, n. The relevant equations are:

$$\bar{R} = \sigma d_1 \quad (4.4)$$

$$\text{Lower warning line} = \bar{R}w_1 \quad (4.5)$$

$$\text{Upper warning line} = \bar{R}w_2 \quad (4.6)$$

$$\text{Lower action line} = \bar{R}a_1 \quad (4.7)$$

$$\text{Upper action line} = \bar{R}a_2 \quad (4.8)$$

EXAMPLE 4.7.1

Determine the characteristics of the mean and range control charts for a process in which the target value is 57, the process capability is 5, and the sample size is 4.

For the control chart on which mean values will be plotted, the calculation is simple. The warning lines will be at $57 \pm 2 \times 5/\sqrt{4}$, i.e. at 57 ± 5; and the action lines will be at $57 \pm 3 \times 5/\sqrt{4}$, i.e. at 57 ± 7.5. This chart is shown in Figure 4.3a.

For the control chart on which ranges are plotted, we must first calculate $\bar{R}$ using equation (4.4). This gives $\bar{R} = 5 \times 2.059 = 10.29$, where the d_1 value of 2.059 is taken from statistical tables for $n = 4$. (See for example the table in the collection by Neave, the details of which are given in the Bibliography for Chapter 1.) The value of $\bar{R}$ is then used to determine the lower and upper warning and action lines using equations (4.5)–(4.8). The values of w_1, w_2, a_1 and a_2 for $n = 4$ are 0.29, 1.94, 0.10, and 2.58 respectively, giving on multiplication by 10.29 positions for the four lines of 2.98, 19.96, 1.03 and 26.55 respectively. These lines are shown in Figure 4.3b.

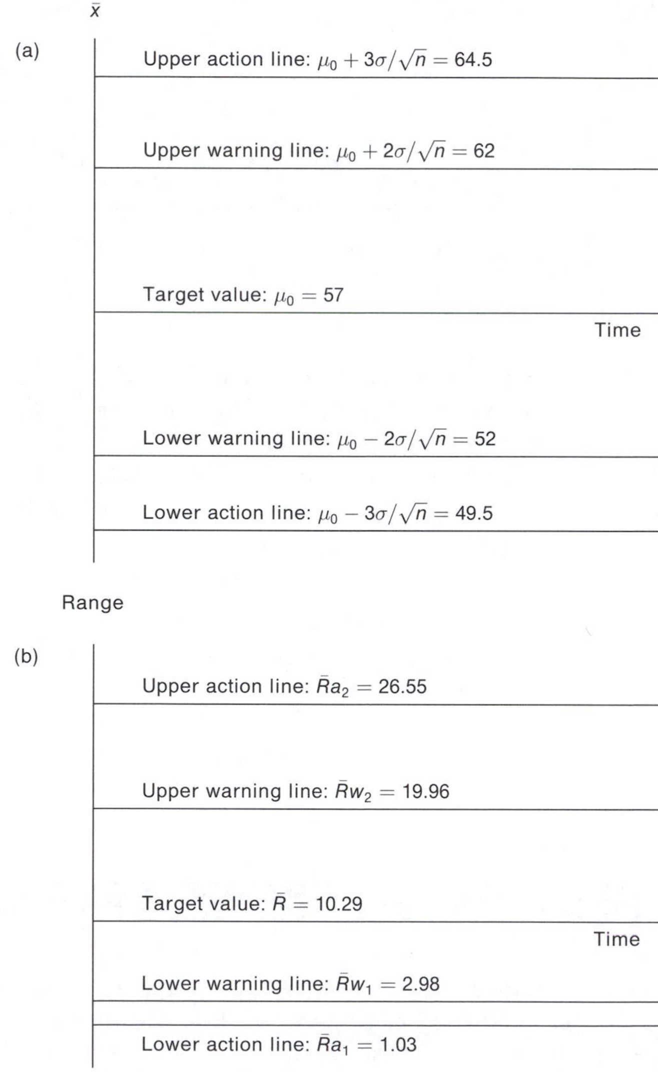

Figure 4.3 (a) Shewhart chart for mean values (example). (b) Shewhart chart for range (example).

It is not always the practice to plot the lower action and warning lines on a control chart for the range, as a reduction in the range is not normally a cause for concern. However, as already noted, the variability of a process is one measure of its quality, and a reduction in $\bar{R}$ represents an improvement in quality, the causes of which may be well worth investigating. So plotting both sets of warning and action lines is recommended.

4.8 Establishing the process capability

In the previous section we showed that, if the process capability, σ, is known, it is possible to construct control charts for both the sample mean and the sample range. It is thus possible to distinguish between a situation where a process has gone out of control by a shift in the process mean from a situation where the mean is unchanged but an undesirable increase in the variability of the process has occurred. The establishment of a proper value for σ is therefore very important, and such a value should be based on a substantial number of measurements. But in making such measurements the same problem – distinguishing a change in the process mean from a change in the process variability – must be faced. If σ is calculated directly from a long sequence of measurements, its value may be over-estimated by any changes in the mean that occur during that sequence, and proper control charts could not then be plotted.

The solution to this problem is to take a large number of small samples, measure the range, R, for each, and thus determine $\bar{R}$. This procedure ensures that only the inherent variability of the process is measured, with any drift in the mean values eliminated. The $\bar{R}$-value can then be used with equations (4.5)–(4.8) to determine the action and warning lines for the range control chart. The warning and action lines for the control chart for the mean can be determined by calculating σ using equation (4.4), and then applying equations (4.2) and (4.3). In practice this two-stage calculation is unnecessary, as most statistical tables provide values of W and A, which give the positions of the warning and action lines directly from:

$$\text{Warning lines at } \bar{x} \pm W\bar{R} \tag{4.9}$$

$$\text{Action lines at } \bar{x} \pm A\bar{R} \tag{4.10}$$

These methods are illustrated by the following example.

EXAMPLE 4.8.1

An internal quality control standard with an analyte concentration of 50 mg kg^{-1} is analysed in a laboratory for 25 consecutive days, the sample size being four on each day. The results are given in Table 4.2, which is in the form of an Excel spreadsheet. Determine the value of $\bar{R}$ and hence plot control charts for the mean and range of the laboratory analyses.

When the results are examined there is clearly some evidence that, over the 25-day period of the analyses, the sample means are drifting up and down. All the sample means from days 3–15 inclusive are greater than the target value of 50, whereas four of the next six means are below the target value, and the last four are all above it. These are the circumstances in which it is important to estimate σ using the method described above. Using the R-values in the last column of data, $\bar{R}$ is found to be 4.31. Application of equation (4.4) estimates σ as $4.31/2.059 = 2.09$. Table 4.2 also shows that the standard deviation of the 100 measurements, treated as a single sample, is 2.43: because of the drifts in the mean this would be a significant overestimate of σ.

The control chart for the mean is then plotted with the aid of equations (4.9) and (4.10) with $W = 0.4760$, $A = 0.7505$, which show that the warning and action lines are at 50 ± 2.05 and 50 ± 3.23 respectively. Figure 4.4 shows the Excel control chart. Similarly, equations (4.5)–(4.8) show that in the control chart for the range the warning lines are at 1.24 and 8.32 and the action lines are at 0.42 and 11.09 respectively. (Excel does not automatically produce control charts for ranges, though it does generate charts for standard deviations, which in some cases are used instead of range charts.) Figure 4.4 shows that this process is not yet in control, several of the points falling outside the (upper) action line.

Table 4.2 Excel spreadsheet (example)

Sample Number	Sample Values 1	2	3	4	Chart Mean		Range
1	48.8	50.8	51.3	47.9	49.70		3.4
2	48.6	50.6	49.3	50.3	49.70		2.0
3	48.2	51.0	49.3	52.1	50.15		3.9
4	54.8	54.6	50.7	53.9	53.50		4.1
5	49.6	54.2	48.3	50.5	50.65		5.9
6	54.8	54.8	52.3	52.5	53.60		2.5
7	49.0	49.4	52.3	51.3	50.50		3.3
8	52.0	49.4	49.7	53.9	51.25		4.5
9	51.0	52.8	49.7	50.5	51.00		3.1
10	51.2	53.4	52.3	50.3	51.80		3.1
11	52.0	54.2	49.9	57.1	53.30		7.2
12	54.6	53.8	51.5	47.9	51.95		6.7
13	52.0	51.7	53.7	56.8	53.55		5.1
14	50.6	50.9	53.9	56.0	52.85		5.4
15	54.2	54.9	52.7	52.2	53.50		2.7
16	48.0	50.3	47.5	53.4	49.80		5.9
17	47.8	51.9	54.3	49.4	50.85		6.5
18	49.4	46.5	47.7	50.8	48.60		4.3
19	48.0	52.5	47.9	53.0	50.35		5.1
20	48.8	47.7	50.5	52.2	49.80		4.5
21	46.6	48.9	50.1	47.4	48.25		3.5
22	54.6	51.1	51.5	54.6	52.95		3.5
23	52.2	52.5	52.9	51.8	52.35		1.1
24	50.8	51.6	49.1	52.3	50.95		3.2
25	53.0	46.6	53.9	48.1	50.40		7.3
			s.d. =	2.43		Mean =	4.31

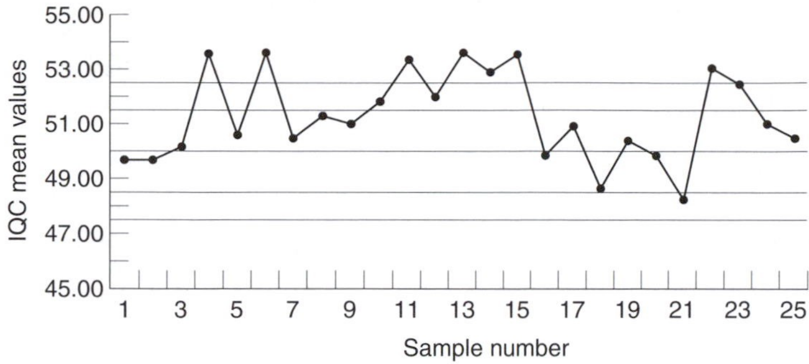

Figure 4.4 Shewhart chart for means (Table 4.2 example data).

4.9 Average run length: cusum charts

An important property of a control chart is the speed with which it detects that a change in the process mean has occurred. The average number of measurements necessary to detect any particular change in the process mean is called the **average run length** (ARL). Since the positions of the action and warning lines on a Shewhart chart for the process mean depend on the value of $\sigma/\sqrt{n}$, the ARL for that chart will depend on the size of the change in the mean compared with $\sigma/\sqrt{n}$. A larger change will be detected more rapidly than a smaller one, and the ARL will be reduced by using a larger sample size, n. It may be shown that if a change equal to $1\sigma/\sqrt{n}$ occurs, then the ARL is about 50 if only the action line criterion is used, i.e. about 50 samples will be measured before a value falls outside the action lines. If the process is also stopped if two consecutive measurements fall outside the same warning line, then the ARL falls to ca. 25. These values are quite large: for example it would be serious if a laboratory continued a pesticide analysis for 25 days before noticing that the procedure had developed a systematic error. This represents a significant disadvantage of Shewhart charts. An example of the problem is shown in Table 4.3, a series of measurements for which the target value is 80, and $\sigma/\sqrt{n}$ is 2.5. When the sample means are plotted on a Shewhart chart (Figure 4.5) it is clear that from about the seventh observation onwards a change in the process mean may well have occurred, but all the points remain on or inside the warning lines. (Only the lower warning and action lines are shown in the figure.)

The ARL can be reduced significantly by using a different type of control chart, a **cusum (cumulative sum)** chart. This approach is again illustrated by the data in Table 4.3. The calculation of the cusum is shown in the last two columns of the table, which show that the sum of the deviations of the sample means from the target value is carried forward cumulatively, careful attention being paid to the signs of the deviations. If a manufacturing or analytical process is under control, positive and negative deviations from the target value are equally likely and the cusum should oscillate about zero. If the process mean changes the cusum will move away from zero. In the example

Table 4.3 Example data for cusum calculation

Observation number	*Sample mean*	*Sample mean – target value*	*Cusum*
1	82	2	2
2	79	−1	1
3	80	0	1
4	78	−2	−1
5	82	2	1
6	79	−1	0
7	80	0	0
8	79	−1	−1
9	78	−2	−3
10	80	0	−3
11	76	−4	−7
12	77	−3	−10
13	76	−4	−14
14	76	−4	−18
15	75	−5	−23

given, the process mean seems to fall after the seventh observation, so the cusum becomes more and more negative. The resulting control chart is shown in Figure 4.6.

Proper interpretation of cusum charts, to show that a genuine change in the process mean has occurred, requires a **V-mask**. The mask is engraved on a transparent plastic sheet, and is placed over the control chart with its axis of symmetry horizontal and its apex a distance, d, to the right of the last observation (Figure 4.7). If all the points on the chart lie within the arms of the V, then the process is in control (Figure 4.7a). The mask is also characterized by $\tan\theta$, the tangent of the semi-angle, θ, between the arms of the V. Values of d and $\tan\theta$ are chosen so that significant changes in the process mean are

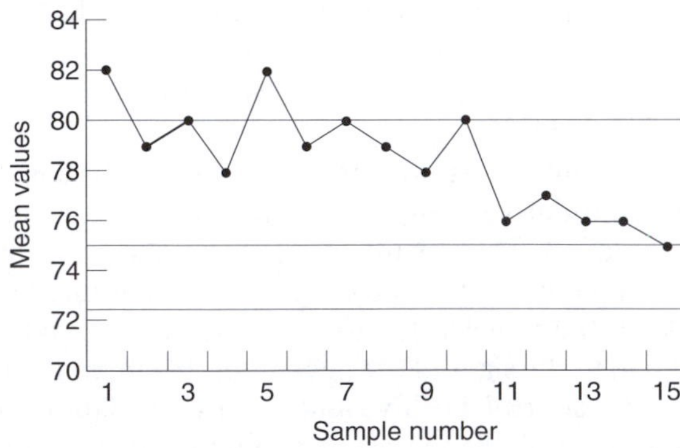

Figure 4.5 Shewhart chart for Table 4.3 data.

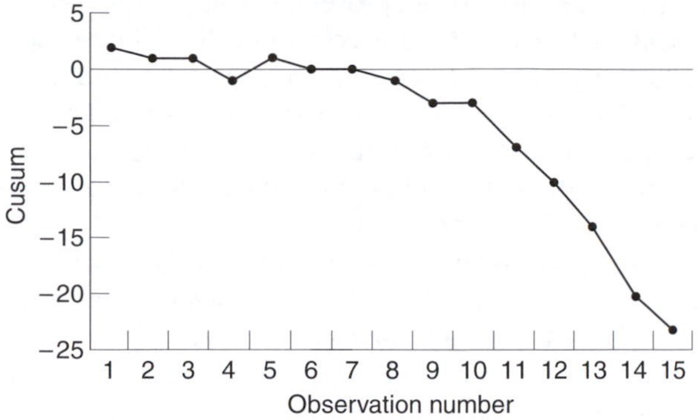

Figure 4.6 Cusum chart for Table 4.3 data.

detected quickly, but false alarms are few. The unit of d is the distance between successive observations. The value of $\tan\theta$ used clearly depends on the relative scales of the two axes on the chart: a commonly used convention is to make the distance between successive observations on the x-axis equal to $2\sigma/\sqrt{n}$ on the y-axis. A V-mask with $d = 5$ units and $\tan\theta = 0.35$ then gives an ARL of 10 if the process mean changes by $1\sigma/\sqrt{n}$ and only 4 if the

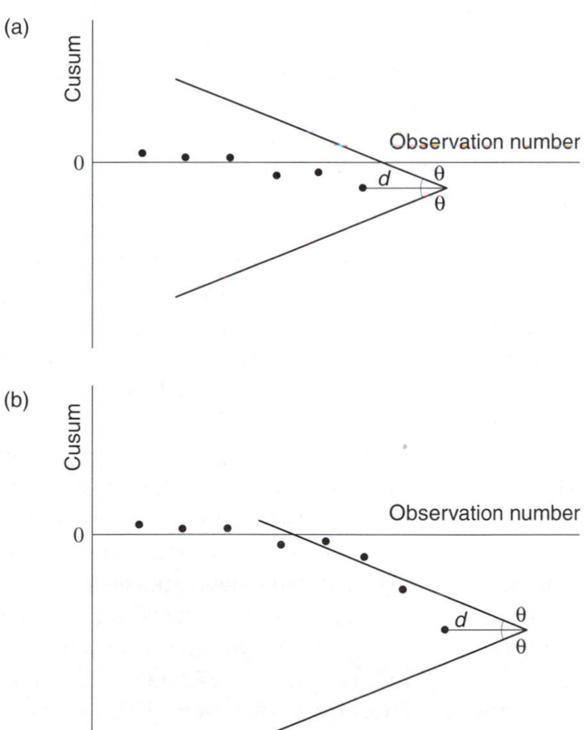

Figure 4.7 (a) Use of a V-mask with the process in control. (b) Use of a V-mask with the process out of control.

change is $2\sigma/\sqrt{n}$. The ARL for a zero change in process mean, i.e. before a false alarm occurs, is ca. 350. The corresponding figures for a Shewhart chart are ca. 25 (for a change in the mean of $1\sigma/\sqrt{n}$) and 320, so it is clear that the cusum chart is superior in both respects. The ARL provided by the cusum chart can be reduced to about 8 (for a change of $1\sigma/\sqrt{n}$) by using $\tan\theta = 0.30$, but inevitably the chance of a false alarm is then also increased, occurring once in ca. 120 observations.

In summary, cusum charts have the advantage that they react more quickly than Shewhart charts to a change in the process mean (as Figure 4.6 clearly shows), without increasing the chances of a false alarm. The point of the slope change in a cusum chart indicates the point where the process mean has changed, and the value of the slope indicates the size of the change. Naturally, if a cusum chart suggests that a change in the process mean has occurred, we must also test for possible changes in σ. This can be done using a Shewhart chart, but cusum charts for ranges can also be plotted. More details on these uses of control charts are given in the references at the end of the chapter.

4.10 Proficiency testing schemes

The quality of analytical measurements is enhanced by two types of testing scheme, in each of which a number of laboratories might participate simultaneously. In the first of these, **proficiency testing (PT) schemes**, aliquots from homogeneous materials are circulated to a number of laboratories for analysis at regular intervals (every few weeks or months), and the resulting data are reported to a central organizer. Each laboratory analyses its portion *using its own usual method*, and the material circulated is designed to resemble as closely as possible the samples normally submitted for analysis in the relevant field of application. The results of all the analyses are circulated to all the participants, who thus gain information on how their measurements compare with those of others, how their own measurements improve or deteriorate with time, and how their own measurements compare with an external quality standard. In short, the aim of such schemes is the evaluation of the competence of analytical laboratories. PT schemes have now been developed for use in a wide range of application fields including several areas of clinical chemistry, water analysis, various types of food and drink, forensic analysis, and so on. Experience shows that in such schemes widely divergent results will arise, even between experienced and well-equipped and -staffed laboratories. In one of the commonest of clinical analyses, the determination of blood glucose at the mM level, most of the results obtained for a single blood sample approximated to a normal distribution with values between 9.5 and 12.5 mM, in itself a not inconsiderable range. But the complete range of results was from 6.0 to 14.5 mM, i.e. some laboratories obtained values almost 2.5 times those of others. The worrying implications of this discrepancy in clinical diagnosis are obvious. In more difficult areas of analysis the results can be so divergent that there is no real consensus between different laboratories. The importance of PT schemes in highlighting such alarming differences, and in helping to minimize them by encouraging laboratories to compare their performance, is very clear, and they have unquestionably helped to improve the quality of analytical results in many fields. Here we are concerned only with the statistical evaluation of the design and results of such schemes, and not with the administrative details of their organization. Of particular importance are the methods of assessing participants' performance and the need to ensure that the bulk sample from which aliquots are circulated is homogeneous.

The recommended method for verifying homogeneity of the sample involves taking $n \geq 10$ portions of the test material at random, separately homogenizing them if necessary, taking two test samples from each portion, and analysing the $2n$ portions by a method whose standard deviation under repeatability conditions is not more than 30% of the target standard deviation (i.e. the expected reproducibility, see below) of the proficiency test. If the homogeneity is satisfactory, one-way analysis of variance should then show that the between-sample mean square is not significantly greater than the within-sample mean square (see Section 4.3).

The results obtained by the laboratories participating in a PT scheme are most commonly expressed as z-scores, where z is given by (see Section 2.2):

$$z = \frac{x - x_a}{\sigma} \tag{4.11}$$

In this equation the x value is the result obtained by a single laboratory for a given analysis; x_a is the assigned value for the level of the analyte, and σ is the target value for the standard deviation of the test results. The assigned value x_a is best obtained by using a certified reference material, if one is available and suitable for distribution to the participants. In some cases this is not possible, and the assigned value is the mean value obtained by a number of selected 'expert' laboratories. In still other cases the only feasible assigned value is a consensus obtained from the results of most or all of the laboratories. This last situation is of interest since, when many laboratories participate in a given PT scheme, there are bound to be a number of suspect results or outliers in an individual test. (It should be noted that, although many PT schemes provide samples and reporting facilities for more than one analyte, experience shows that a laboratory that scores well in one specific analysis does not necessarily score well in others.) This problem is overcome either by the use of the *median* (see Chapter 6), which is especially recommended for small data sets ($n < 10$), a *robust mean* (see Chapter 6), or the *mean of the inter-quartile range* (see Chapter 6). All these measures of location avoid or address the effects of dubious results. It is also recommended that the *uncertainty* of the assigned value is reported to participants in the PT. This also may be obtained from the results from expert laboratories: estimates of uncertainty are covered in more detail below (Section 4.12).

The target value for the standard deviation, σ, should be circulated in advance to the PT participants along with a summary of the method by

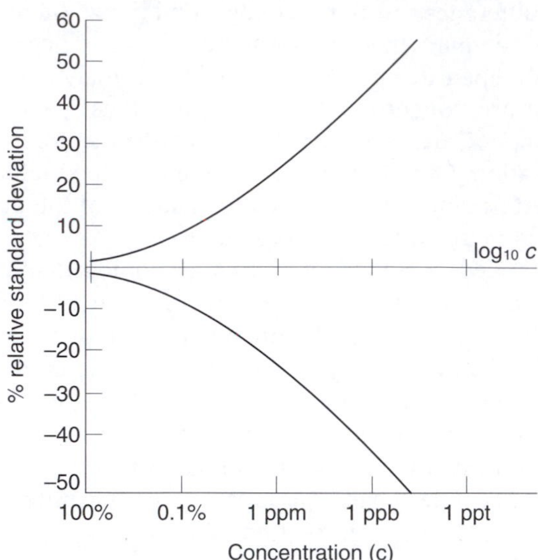

Figure 4.8 The Horwitz trumpet.

which it has been established. It will vary with analyte concentration, and one approach to estimating it is to use a functional relationship between concentration and standard deviation. The best known relationship is the Horwitz trumpet, dating from 1982, so called because of its shape. Using many results from collaborative trials, Horwitz showed that the relative standard deviation of a method varied with the concentration, c (e.g. mg g^{-1}), according to the approximate and empirical equation:

$$\mathrm{RSD} = \pm 2^{(1-0.5\log c)} \tag{4.12}$$

This equation leads to the trumpet-shaped curve shown in Figure 4.8, which can be used to derive target values of σ for any analysis. Such target values can also be estimated from prior knowledge of the standard deviations usually achieved in the analysis in question. Another approach uses fitness for purpose criteria: if the results of the analysis, used routinely, require a certain precision for the data to be interpreted properly and usefully, that precision provides the largest (worst) acceptable value of σ. It is poor practice to estimate σ from the results of previous rounds of the PT scheme itself, as this would conceal any improvement or deterioration in the quality of the results with time.

The results of a single round of a PT scheme are frequently summarized as shown in Figure 4.9. If the results follow a normal distribution with mean x_a and standard deviation σ, the z-scores will be a sample from the standard normal distribution, i.e. a normal distribution with mean zero and variance 1. Thus a laboratory with a $|z|$ value of <2 is generally regarded as

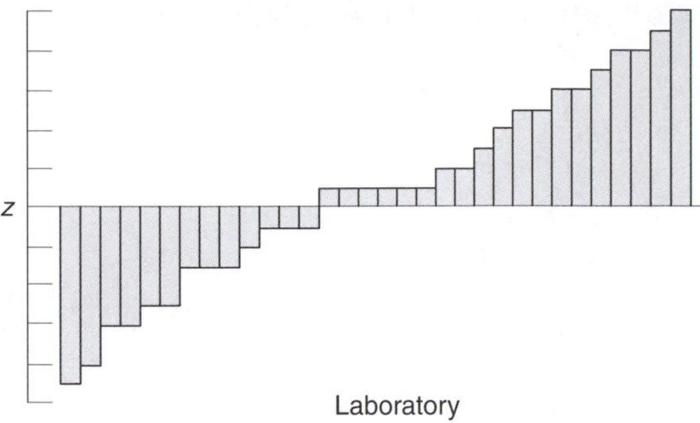

Figure 4.9 Summary of results of a single PT round.

having performed satisfactorily, a $|z|$ value between 2 and 3 is questionable, and $|z|$ values >3 are unacceptable. Of course even the laboratories with satisfactory scores will strive to improve their values in the subsequent rounds of the PT. In practice it is not uncommon to find 'heavy-tailed' distributions, i.e. more results than expected with $|z| > 2$.

Some value has been attached to methods of combining z-scores. For example the results of one laboratory in a single PT scheme over a single year might be combined (though this would mask any improvement or deterioration in performance over the year). If the same analytical method is applied to different concentrations of the same analyte in each round of the same PT scheme, again a composite score might have limited value. Two functions used for this purpose are the **re-scaled sum of z-scores** (RSZ), and the **sum of squared z-scores** (SSZ), given by $\text{RSZ} = \sum_i z_i/\sqrt{n}$ and $\text{SSZ} = \sum_i z_i^2$ respectively. Each of these functions has disadvantages, and the use of combined z-scores is not to be recommended.

4.11 Collaborative trials

As we have seen, proficiency testing schemes allow the *competence of laboratories* to be monitored, compared and perhaps improved. By contrast a **collaborative trial** (CT) aims to evaluate the precision of an *analytical method*, and sometimes its ability to provide results free from bias. It is normally a one-off experiment involving expert or competent laboratories, all of which by definition use the same technique. Collaborative trials are perhaps better described as **method performance studies**.

A crucial preliminary experiment is the 'ruggedness' test. As we saw in Chapter 1, even very simple analytical experiments involve several individual steps and perhaps the use of a number of reagents. Thus many experimental factors (e.g. temperature, solvent composition, pH, humidity, reagent purity, concentration, etc.) will affect the results, and it is essential that such factors are identified and studied before any collaborative trial is undertaken. In some cases a method is found to be so sensitive to small changes in one

Table 4.4 Ruggedness test for seven factors

Experiment	*Factors*							*Result*
	A	*B*	*C*	*D*	*E*	*F*	*G*	
1	+	+	+	+	+	+	+	y_1
2	+	+	−	+	−	−	−	y_2
3	+	−	+	−	+	−	−	y_3
4	+	−	−	−	−	+	+	y_4
5	−	+	+	−	−	+	−	y_5
6	−	+	−	−	+	−	+	y_6
7	−	−	+	+	−	−	+	y_7
8	−	−	−	+	+	+	−	y_8

factor that is in practice very difficult to control (e.g. very high reagent purity) and the method is rejected as impracticable before a CT takes place. In other instances the trial will continue, but the collaborators will be warned of the factors to be most carefully controlled. Although a more complete discussion of experimental design is deferred to Chapter 7, it is important to indicate here that much information can be obtained from a relatively small number of experiments. Suppose it is believed that seven experimental factors (A–G) might affect the results of an analysis. These factors must be tested at (at least) two values, called levels, to see whether they are really significant. Thus, if temperature is thought to affect the result, we must perform preliminary experiments at two temperatures (levels) and compare the outcomes. Similarly, if reagent purity may be important, experiments with high purity and lower purity reagent batches must be done. It might thus be thought that 2^7 preliminary experiments, covering all the possible combinations of seven factors at two levels, will be necessary. In practice, however, just eight experiments can provide important information. The two levels of the factors are called + and −, and Table 4.4 shows how these levels are set in the eight experiments, the results of which are called y_1, $y_2, \ldots, y_8$. The effect of altering each of the factors from its high level to its low level is easily calculated. Thus the effect of changing B from + to − is $(y_1 + y_2 + y_5 + y_6)/4 - (y_3 + y_4 + y_7 + y_8)/4$.

When the seven differences for factors A–G have all been calculated in this way, it is easy to identify any factors that have a worryingly large effect on the results. It may be shown that any difference that is more than twice the standard deviation of replicate measurements is significant and should be further studied. This simple set of experiments, technically known as an **incomplete factorial design**, has the disadvantage that interactions between the factors cannot be detected. This point is further discussed in Chapter 7.

In recent years international bodies have moved towards an agreement on how CTs should be performed. At least eight laboratories ($k \geq 8$) should be involved. Since the precision of a method usually depends on the analyte concentration it should be applied to at least five different levels of analyte in the same sample matrix with duplicate measurements ($n = 2$) at each level. A crucial requirement of a CT is that it should distinguish between the repeatability standard deviation, s_r, and the reproducibility standard deviation, s_R. At each analyte level these are related by the equation:

$$s_R^2 = s_r^2 + s_L^2 \qquad (4.13)$$

where s_L^2 is the variance due to inter-laboratory differences, which reflect different degrees of bias in different laboratories. Note that in this particular context, reproducibility refers to errors arising in different laboratories and equipment, but using the *same* analytical method: this is a more restricted definition of reproducibility than that used in other instances. As we saw in Section 4.3, one-way analysis of variance can be used (with separate calculations at each concentration level used in the CT) to separate the sources of variance in equation (4.13). However, the proper use of the equation involves two assumptions: (1) that at each concentration level the means obtained in different laboratories are normally distributed; and (2) that at each concentration the repeatability variance among laboratories is equal. Both these assumptions are tested using standard methods before the ANOVA calculations begin. In practice the second assumption, that of homogeneity of variance, is tested first using Cochran's method. Strictly speaking this test is designed to detect outlying variances rather than testing for homogeneity of variance as a whole, but other more rigorous methods for the latter purpose are also more complex. Cochran's test calculates C by comparing the largest range (i.e. difference between the two results from a single laboratory) with the sum of all such ranges. (If $n > 2$ variances rather than ranges are compared, but here we assume that each participating laboratory makes just two measurements at each level):

$$C = \frac{w_{\max}^2}{\sum_j w_j^2} \qquad (4.14)$$

where j takes values from 1 to k, the number of participating laboratories. The value of C obtained is compared with the critical values in Table A.15, and the null hypothesis, i.e. that the largest variance is not an outlier, is rejected if the critical value at the appropriate value of k is exceeded. When the null hypothesis is rejected, the results from the laboratory in question are discarded.

The first assumption is then tested using Grubbs' test (Section 3.7) which is applied first as a test for single outliers, and then (since each laboratory makes duplicate measurements) in a modified form as a test for paired outliers. In

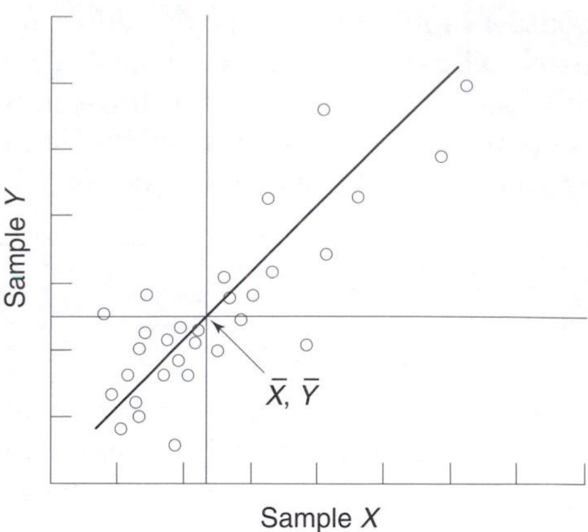

Figure 4.10 A Youden two-sample plot.

both cases all the results from laboratories producing outlying results are again dropped from the trial unless this would result in the loss of too much data. When these outlier tests are complete, the ANOVA calculation can proceed as in Section 4.3.

In many circumstances it is not possible to carry out a full CT as described above, for example when the test materials are not available with a suitable range of analyte concentrations. In such cases a simpler system can be used. This is the **Youden matched pairs** or two-sample method, in which each participating laboratory is sent *two materials of similar composition*, X and Y, and asked to make one determination on each. The results are plotted as shown in Figure 4.10, each point on the plot representing a pair of results from one laboratory. The mean values for the two materials, $\bar{X}$ and $\bar{Y}$, are also determined, and vertical and horizontal lines are drawn through the point $(\bar{X}, \bar{Y})$, thus dividing the chart into four quadrants. This plot allows us to assess the occurrence of random errors and bias in the trial. If only random errors occur the X and Y determinations may give results which are both too high, both too low, X high and Y low, or X low and Y high. These four outcomes would be equally likely, and the number of points in each of the quadrants would be roughly equal. But if a systematic error occurs in a laboratory, it is likely that its results for both X and Y will be high, or both will be low. So if systematic errors dominate, most of the points will be in the top-right and bottom-left quadrants. This is indeed the result that is obtained in most cases. In the impossible event that random errors were absent, all the results would lie on a line at 45° to the axes of the plot, so when in practice such errors do occur, the perpendicular distance of a point from that line is a measure of the random error of the laboratory. Moreover the distance from the intersection of that perpendicular with the 45° line to the point $(\bar{X}, \bar{Y})$ measures the systematic error of the laboratory. This fairly simple approach to a collaborative trial is thus capable of yielding a good deal of information in a simple form. The Youden approach has the further advantages that participating laboratories are not tempted to censor one or more replicate determinations, and that more materials can be studied without large numbers of experiments.

Youden plots provide a good deal of information in an immediately accessible form, but we still need methods for calculating the variances s_R^2 and s_r^2. The following example shows how this can also be done in a simple way.

EXAMPLE 4.11.1

The lead levels (in ng g^{-1}) in two similar samples (X and Y) of solid milk formulations for infants were determined in nine laboratories (1–9) by graphite-furnace atomic-absorption spectrometry. The results were:

Sample	*Laboratories*								
	1	*2*	*3*	*4*	*5*	*6*	*7*	*8*	*9*
X	35.1	23.0	23.8	25.6	23.7	21.0	23.0	26.5	21.4
Y	33.0	23.2	22.3	24.1	23.6	23.1	21.0	25.6	25.0

Evaluate the overall interlaboratory variation, and its random and systematic components.

In CTs of this type there is a difference between the samples as well as the differences between laboratories. In the normal way, such a situation would be dealt with by two-way ANOVA (see Section 7.4), and in some cases this is done. However, in this instance there are only two samples, deliberately chosen to be similar in their analyte content, so there is little interest in evaluating the difference between them. The calculation can therefore be set out in a way that is numerically and conceptually simpler than a complete two-way ANOVA. In performing the calculation we know that the result obtained by each laboratory for sample X may include a systematic error. The *same* systematic error will presumably be included in that laboratory's result for the similar sample Y. The difference D $(= X - Y)$ will thus have this error removed, so the spread of the D values will provide an estimate of the random or measurement errors. Similarly, X and Y can be added to give T, the spread of which gives an estimate of the overall variation in the results. The measurement variance is then estimated by:

$$s_r^2 = \frac{\sum_i (D_i - \bar{D})^2}{2(n-1)} \tag{4.15}$$

and the overall variance, s_R^2, due to all sources of error, is estimated by:

$$s_R^2 = \frac{\sum_i (T_i - \bar{T})^2}{2(n-1)} \tag{4.16}$$

Notice that each of these equations includes a 2 in the denominator. This is because D and T each give estimates of errors in two sets of results, subtracted and added in D and T respectively. The results of this trial can be expressed in a table as follows:

	1	2	3	4	5	6	7	8	9
X	35.1	23.0	23.8	25.6	23.7	21.0	23.0	26.5	21.4
Y	33.0	23.2	22.3	24.1	23.6	23.1	21.0	25.6	25.0
D	2.1	−0.2	1.5	1.5	0.1	−2.1	2.0	0.9	−3.6
T	68.1	46.2	46.1	49.7	47.3	44.1	44.0	52.1	46.4

The third and fourth rows of the table can be used to show that $\bar{D} = 0.244$ and $\bar{T} = 49.33$. Equations (4.15) and (4.16) then show that the overall variance and the measurement variances are $(5.296)^2$ and $(1.383)^2$ respectively. These can be compared as usual using the F-test, giving $F = 14.67$. The critical value, $F_{8,8}$, is 3.44 ($P = 0.05$), so the interlaboratory variation cannot simply be accounted for by random errors. The component due to bias, s_L^2, is given here by

$$s_R^2 = 2s_L^2 + s_r^2 \tag{4.17}$$

Note again the appearance of the 2 in equation (4.17), because two sample materials are studied. Here it is a simple matter to calculate that the estimate of s_L^2 is $(3.615)^2$. The mean of all the measurements is $49.33/2 = 24.665$, so the relative standard deviation is $(100 \times 5.296)/24.665 = 21.47\%$. This seems to be a high value, but the Horwitz trumpet relationship would predict an even higher value of ca. 28% at this concentration level. It should be noted that possible outliers are not considered in the Youden procedure, so the possibility of rejecting the results from laboratory 1 does not arise.

4.12 Uncertainty

In Chapter 1 we learned that analytical procedures will be affected by both random errors and by bias. In fairly recent years analytical chemists have increasingly recognized the importance of providing for each analysis a single number which describes their combined effect. The **uncertainty** of a result is a parameter that describes a range within which the value of the quantity being measured is expected to lie, taking into account all sources of error. The concept is well established in physical measurements. Its value in analytical chemistry is also undeniable, though questions and controversies remain over the ease of its interpretation by legal and statutory bodies and lay people, and about the best methods of calculating it. Two symbols are used to express uncertainty. **Standard uncertainty** (u) expresses the concept as a standard deviation. **Expanded uncertainty** (U) defines a *range* that encompasses a large fraction of the values within which the quantity being measured will lie and is obtained by multiplying u by a **coverage factor**, k, chosen according to the degree of confidence required for the range, i.e. $U = u \times k$. Since u is analogous to a standard deviation, if k is 2 (this is generally taken as the default value if no other information is given), then U gives approximately one-half of the 95% confidence interval.

In principle, two basic approaches to estimating uncertainty are available. The **bottom-up** approach identifies each separate stage of an analysis, including sampling steps wherever possible, assigns appropriate random and systematic errors to each, and then combines these components using the rules summarized in Section 2.11 to give an overall u value. However for a number of reasons this process may not be as simple as it seems. The first problem is that even simple analytical processes may involve many individual experimental steps and possible sources of error. It is easy to overlook some of these sources and thus arrive at an over-optimistic uncertainty value. If all the sources of error *are* all fully identified, then the whole calculation process is liable to be long-winded. Examples of error sources that should be considered but are easily overlooked include operator bias; instrument bias, including sample carry-over; assumptions concerning reagent purity; use of volumetric apparatus at a temperature different from that at which it was calibrated; changes in the composition of the sample during the analysis, either because of contamination or because of inherent instability; use of calculators or computers with inadequate capabilities or with the wrong statistical model applied; and so on. All these factors may arise *in addition* to the random errors that inevitably occur in repeated measurements. Whereas the latter may be estimated directly by repeated measurements, some of the former may not be amenable to experiment, and may have to be estimated using experience, or equipment manufacturers' information such as calibration certificates or instrument specifications.

Another problem is that, as shown in Chapter 2, systematic errors do not immediately lend themselves to statistical treatment in the same way as random errors. How then can they be combined with random errors to give an overall u value? (It is still good practice to minimize systematic errors by the use of standards and reference materials, but we should still include the errors involved in that correction process in the overall uncertainty estimate.) The usual method of tackling systematic errors is to treat them as coming from a rectangular distribution. Suppose for example that a manufacturer quotes the purity of a reagent as $99.9 \pm 0.1\%$. This does not mean that the purity of the reagent in its container varies randomly with a standard deviation of 0.1%: it means that the purity of the reagent in a single bottle is between 99.8% and 100.0%. That is, any single bottle provides a systematic error, and there is no reason to suppose that the actual purity is closer to 99.9% than to any other value in the range 99.8–100.0%. In such cases, the contribution to the standard uncertainty is obtained by dividing the error by $\sqrt{3}$, giving a value of $0.1/\sqrt{3} = 0.0577$, and combining this value with other contributions as if it came from a source of random error.

A further problem, the extent of which seems not to have been fully investigated, is that the rules for combining errors given in Chapter 2 assume that the sources of the errors are *independent*. In reality it seems quite possible that this is not always true. For example if a series of experiments is conducted over a period in which the laboratory temperature fluctuates, such fluctuations might have several effects, such as altering the capacity of volumetric apparatus, causing sample losses through volatility, affecting the sensitivity of optical or electrochemical detectors, and so on. Since all these errors would arise from a single source, they would be correlated, and strictly speaking could not be combined using the simple formulae. In such cases the actual uncertainty might be less than the u value calculated on the assumption of independent errors.

Overall the bottom-up approach to uncertainty estimates may be too time-consuming for many purposes. It is possible that in some laboratories it is not necessary to make such calculations very often, as an uncertainty estimate made in detail for one analysis may serve as a model for other analyses over a period of time. But in other instances, most obviously where legal or regulatory issues arise (see below), this will not be sufficient and an uncertainty estimate will have to be provided for each disputed sample. Despite this the bottom-up approach is the one currently recommended by many authorities.

A completely different approach is the **top-down** method, which seeks to use the results of proficiency testing schemes in a number of laboratories (see Section 4.10) to give estimates of the overall uncertainties of the measurements without necessarily trying to identify every individual source of error. The method is clearly only applicable in areas where data from properly run proficiency schemes are available, though such schemes are rapidly expanding in number and may thus provide a real alternative to bottom-up methods in many fields. It can be argued that the uncertainty values calculated in this way are more realistic than bottom-up values, and there is a great saving of effort, since the PT scheme results provide uncertainty estimates directly. On the other hand, PT schemes use a variety of analytical methods, so it might reasonably be claimed that the uncertainty of results from a laboratory that has long experience of a single method might be better (smaller) than PT results would suggest. Again, PT schemes utilize single sample materials prepared with great care. Some sampling errors that would occur in a genuine analysis might thus be overlooked.

These problems have led some bodies to propose simpler methods, explicitly designed to minimize the workload in laboratories that use a range of analytical procedures. In one such approach the basis principles are: (1) Systematic errors are not included in the uncertainty estimates, but are assessed using reference materials as usual and thus corrected or eliminated. (2) At least 10 replicate measurements are made on stable and well-characterized authentic samples or reference materials. (This again implies that sampling uncertainties are not included in the estimates.) (3) Uncertainties are calculated from the standard deviations of measurements made in *internal reproducibility conditions*, i.e. with different analysts, using different concentrations (including any that are relevant to legal requirements), and in all relevant matrices. These conditions are supposed to mimic those that would arise in a laboratory in everyday operation. Some provision is made

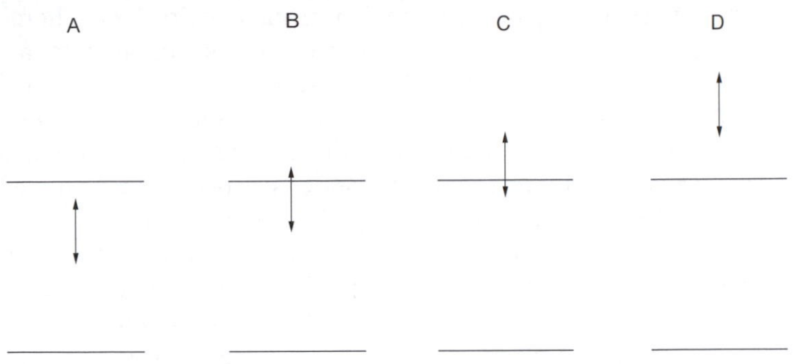

Figure 4.11 Use of uncertainty to test compliance with specification limits.

for circumstances where the reproducibility conditions cannot be achieved (for example where samples are intrinsically unstable). This method seems to be very simple, but it may be adequate: indeed it may be the only practicable method in some instances.

Uncertainty estimates are important not only to anyone who has provided a sample for analysis and who requires a range of values in which the true analyte concentration should lie. They also have value in demonstrating that a laboratory has the capacity to perform analyses of legal or statutory significance. Once an uncertainty value for a particular analysis in a given laboratory is known, it is simple to interpret the results in relation to such statutory or other specification limits. Four possible situations are shown in Figure 4.11, where it is assumed that a coverage factor of 2 has been used to determine U at the 95% level (the 95% interval is shown by the vertical double arrows), and where both upper and lower limits for the concentration of the analyte have been specified. These limits are indicated by the horizontal lines.

In case A the uncertainty interval lies completely between the upper and lower specified limits, so compliance with the specification has been achieved. In case B the 95% interval extends just beyond the upper limit, so although compliance is more likely than not, it cannot be fully verified at the 95% level. In case C compliance is very unlikely, though not impossible, and in case D there is a clear failure to comply.

Although none of the approaches to estimating uncertainties is ideal, and although the term itself still provokes controversy (some analytical chemists think it is too negative or pessimistic in its implications for the lay public), uncertainty calculations seem certain to increase in importance in the future.

4.13 Acceptance sampling

Previous sections of this chapter have shown how the quality of the analytical results obtained in a laboratory can be monitored by internal quality control procedures and by participation in proficiency testing schemes. We have also shown how the concept of uncertainty is designed to help the interpretation of analytical results by the customers for analytical measurements, including regulatory authorities. In this section we consider a further important problem involving both analysts and their customers, called **acceptance sampling**. The simple statistical principles involved have been discussed in previous chapters. Suppose that the manufacturer of a chemical is required to ensure that it does not contain more than a certain level of a particular impurity. This is called the **acceptable quality level** (AQL) of the product and is given the symbol μ_0. The manufacturer's intention to ensure that this impurity level is not exceeded is monitored by testing batches of the product. Each test involves n test portions, whose mean impurity level is found to be $\bar{x}$. The variation between portions, σ, is as we have seen normally known from previous experience. The practical problem that arises is that, even when a batch of manufactured material has an impurity level of μ_0, and is thus satisfactory, values of $\bar{x}$ greater than μ_0 will be found in 50% of the analyses. Therefore the manufacturer establishes a critical value for $\bar{x}$, given the symbol $\bar{x}_0$. For a measured value of $\bar{x} > \bar{x}_0$ the batch is rejected. This critical value is higher than μ_0, thus ensuring that the manufacturer runs only a small risk of rejecting a satisfactory batch.

At the same time the customer wishes to minimize the risk of accepting a batch with a mean impurity level greater than μ_0. This can be achieved by setting an agreed **tolerance quality level** (TQL), μ_1, which has a small probability of acceptance. The aim of acceptance sampling is that the critical value $\bar{x}_0$ should minimize the risk to the customer as well as to the manufacturer. At the same time we wish to ensure that n is no larger than necessary. This can be achieved using the properties of the sampling distribution of the mean, given that σ is known.

Suppose the manufacturer accepts a 5% risk of rejecting a batch of the chemical that is in fact satisfactory, i.e. a batch for which $\bar{x} > \bar{x}_0$, even though $\mu = \mu_0$. Then we can write

$$(\bar{x}_0 - \mu_0)/(\sigma/\sqrt{n}) = 1.64 \qquad (4.18)$$

The value 1.64 can be found in Table A.2 (see also Section 2.2). Suppose also that the customer is prepared to accept a 10% risk of accepting a batch with the impurity at the TQL. Then we can similarly write:

$$(\bar{x}_0 - \mu_1)/(\sigma/\sqrt{n}) = -1.28 \qquad (4.19)$$

Since in practice the values of μ_0 and μ_1 will have been agreed in advance, equations (4.17) and (4.18) provide simultaneous equations that can be solved for n and $\bar{x}_0$.

EXAMPLE 4.13.1

Determine n and $\bar{x}_0$ for the case where the AQL and TQL are 1.00 g kg^{-1} and 1.05 g kg^{-1} impurity respectively, the manufacturer's and customer's risks are 5% and 10% respectively, and σ is 0.05 g kg^{-1}.

The solution to this problem involves the use of equations (7.10) and (7.11) with μ_0 and μ_1 taking values 1.00 and 1.05 respectively. By transformation of these equations we can write:

$$n = [(1.64 + 1.28)0.05/(1.05 - 1.00)]^2$$

$$\bar{x}_0 = [(1.64 \times 1.05) + (1.28 \times 1.00)/(1.64 + 1.28)$$

These equations yield $n = 2.92^2 = 8.53$, which is rounded up to a sample size of 9, and $\bar{x}_0 = 3.002/2.92 = 1.028$. Thus a critical value of 1.028% impurity and sample size of 9 will provide both manufacturer and customer with the necessary assurances.

Bibliography

Lawn, R. E., Thompson, M. and Walker, R. F. 1997. *Proficiency Testing in Analytical Chemistry*. Royal Society of Chemistry, London. (A clear and up to date account of PT schemes.)

Massart, D. L., Vandeginste, B. G. M., Buydens, L. M. C., de Jong, S., Lewi, P. J. and Smeyers-Verbeke, J. 1997. *Handbook of Chemometrics and Qualimetrics, Part A*. Elsevier, Amsterdam. (Comprehensive coverage of many quality related topics.)

Montgomery, D. C. 1985. *Introduction to Statistical Quality Control*. Wiley, New York. (Classic general text on QC statistics.)

Pritchard, E. 1995. *Quality in the Analytical Chemistry Laboratory*. Wiley, Chichester. (Clear introduction to general aspects of quality, with discussion of uncertainty.)

Wernimont, G. T. and Spendley, W. 1985. *Use of Statistics to Develop and Evaluate Analytical Methods*. AOAC, Arlington, USA. (Authoritative sequel to Youden and Steiner.)

Youden, W. J. and Steiner, E. H. 1975. *Statistical Manual of the Association of Official Analytical Chemists*. AOAC, Arlington, USA. (Classic text with much emphasis on collaborative studies.)

Exercises

1. Two sampling schemes are proposed for a situation in which it is known, from past experience, that the sampling variance is 10 and the measurement variance 4 (arbitrary units).

Scheme 1: Take five sample increments, blend them and perform a duplicate analysis.

Scheme 2: Take three sample increments and perform a duplicate analysis on each.

Show that the variance of the mean is the same for both schemes.

What ratio of the cost of sampling to the cost of analysis must be exceeded for the second scheme to be the more economical?

2. The data in the table below give the concentrations of albumin measured in the blood serum of one adult. On each of four consecutive days a blood sample was taken and three replicate determinations of the serum albumin concentration were made.

Day	*Albumin concentrations (normalized, arbitrary units)*		
1	63	61	62
2	57	56	56
3	50	46	46
4	57	54	59

Show that the mean concentrations for different days differ significantly. Estimate the variance of the day-to-day variation (i.e. 'sampling variation').

3. In order to estimate the measurement and sampling variances when the halofuginone concentration in chicken liver is determined, four sample increments were taken from different parts of the liver and three replicate measurements were made on each. The following results were obtained (mg kg^{-1}):

Sample	*Replicate measurements*		
A	0.25	0.22	0.23
B	0.22	0.20	0.19
C	0.19	0.21	0.20
D	0.24	0.22	0.22

Verify that the sampling variance is significantly greater than the measurement variance and estimate both variances.

Two possible sampling schemes are proposed:

Scheme 1: Take six sample increments, blend them and make four replicate measurements.

Scheme 2: Take three sample increments and make two replicate measurements on each.

Calculate the total variance of the mean for each scheme.

4. In order to estimate the capability of a process, measurements were made on six samples of size 4 as shown in the table below (in practice at least 25 such samples would be needed). Estimate the process capability, σ. If the target value is 50, calculate the positions of the action and warning lines for the Shewhart charts for the sample mean and the range.

Sample	*Values*			
1	48.8	50.8	51.3	47.9
2	48.6	50.6	49.3	49.7
3	48.2	51.0	49.3	50.3
4	54.8	54.6	50.7	53.9
5	49.6	54.2	48.3	50.5
6	54.8	54.8	52.3	52.5

5. In a collaborative trial, two closely similar samples of oil shale (A and B) were sent to 15 laboratories, each of which performed a single inductively coupled plasma determination of the cadmium level in each sample. The following results were obtained:

Laboratory	*Cd levels (ppm)*	
	A	*B*
1	8.8	10.0
2	3.8	4.7
3	10.1	12.1
4	8.0	11.0
5	5.0	4.7
6	5.2	6.4
7	6.7	8.7
8	9.3	9.6
9	6.9	7.5
10	3.2	2.8
11	9.7	10.4
12	7.2	8.3
13	6.5	6.8
14	9.7	7.2
15	5.0	6.0

Plot the two-sample chart for these data, and comment on the principal source of error in the collaborative trial. Estimate the overall variance, the measurement variance, and the systematic error component of the variance of the results.

6. The target value for a particular analysis is 120. If preliminary trials show that samples of size 5 give an $\bar{R}$ value of 7, set up Shewhart charts for the mean and range for samples of the same size.

7. An internal quality control sample of blood, used for checking the accuracy of blood alcohol determinations, contains 80.0 mg 100 ml^{-1} of ethanol. Successive daily measurements of the alcohol level in the sample were made using four replicates. The precision (process capability) of the method was known to be 0.6 mg 100 ml^{-1}. The following results were obtained:

Day	*Concentration (mg 100 ml^{-1})*
1	79.8
2	80.2
3	79.4
4	80.3
5	80.4
6	80.1
7	80.4
8	80.2
9	80.0
10	79.9
11	79.7
12	79.6
13	79.5
14	79.3
15	79.2
16	79.3
17	79.0
18	79.1
19	79.3
20	79.1

Plot the Shewhart chart for the mean, and the cusum chart, for these results, and comment on the outcomes.

CHAPTER FIVE

Calibration methods in instrumental analysis: regression and correlation

5.1 Introduction: instrumental analysis

Classical or 'wet chemistry' analysis techniques such as titrimetry and gravimetry remain in use in many laboratories and are still widely taught in Analytical Chemistry courses. They provide excellent introductions to the manipulative and other skills required in analytical work, they are ideal for high-precision analyses, especially when small numbers of samples are involved, and they are sometimes necessary for the analysis of standard materials. However, there is no doubt that most analyses are now performed by instrumental methods. Techniques using absorption and emission spectrometry at various wavelengths, many different electrochemical methods, mass spectrometry, gas and liquid chromatography, and thermal and radiochemical methods, probably account for at least 90% of all current analytical work. There are several reasons for this.

Firstly, instrumental methods can perform analyses that are difficult or impossible by classical methods. Whereas the latter can only rarely detect materials at sub-microgram levels, many instrumental methods are astonishingly sensitive. For example, in recent years fluorescence methods have routinely been used to detect single organic molecules in very small volumes of solution. It is normally only possible to determine one analyte at a time by 'wet chemical' methods, but plasma spectrometry can determine ten or more elements simultaneously (and at very low concentrations). Similarly, methods combining high performance liquid chromatography with a spectroscopic detection procedure can identify and determine many components of complex organic mixtures within a few minutes. Furthermore, the concentration range of a particular classical analysis method is usually limited by practical and theoretical considerations. Thus EDTA titrations can be successfully performed with reactant concentrations as low as about 10^{-4} M, but an upper limit (ca. 0.3 M) is set by the solubility of EDTA in water. The useful concentration range is generally only 2–3 orders of magnitude (i.e. powers of ten) for classical methods. In contrast, some instrumental methods are able to determine analyte concentrations over a range of six or more orders of magnitude: this characteristic has important implications for the statistical treatment of the results, as we shall see in the next section.

Secondly, for a large throughput of samples instrumental analysis is generally quicker and often cheaper than the labour-intensive manual methods. In clinical analysis, for example, there is frequently a requirement for the same analyses to be done on scores or even hundreds of whole blood or blood serum/plasma samples every day. Despite the high initial cost of the equipment, such work is generally performed using completely automatic systems. Automation has become such an important feature of analytical chemistry that the ease with which a particular technique can be automated often determines whether or not it is used at all. A typical automatic method may be able to process samples at the rate of 100 per hour or more. The equipment will take a measured volume of sample, dilute it appropriately, conduct one or more reactions with it, and determine and record the concentration of the analyte or of a derivative produced in the reactions. Other areas where the use of automated equipment is now crucial include environmental monitoring and the rapidly growing field of industrial process analysis. Special problems of error estimation will evidently arise in all these applications of automatic analysis: systematic errors, for example, must be identified and corrected as rapidly as possible.

Lastly, modern analytical instruments are almost always interfaced with personal computers to provide sophisticated system control and the storage, treatment (for example the performance of Fourier transforms or calculations of derivative spectra) and reporting of data. Such systems can also evaluate the results statistically, and compare the analytical results with data libraries in order to match spectral and other information. All these facilities are now available from low-cost computers operating at high speeds. Also important is the use of 'intelligent' instruments, which incorporate automatic set-up and fault diagnosis and can perform optimization processes (see Chapter 7).

The statistical procedures used with instrumental analysis methods must provide as always information on the precision and accuracy of the measurements. They must also reflect the technical advantages of such methods, especially their ability to cover a great range of concentrations (including very low concentrations), and to handle many samples rapidly. (In this chapter we shall not cover methods that facilitate the simultaneous determination of more than one analyte. This topic is outlined in Chapter 8.) In practice the results are calculated and the errors evaluated in a particular way that differs from that used when a single measurement is repeated several times.

5.2 Calibration graphs in instrumental analysis

The usual procedure is as follows. The analyst takes a series of materials (normally at least three or four, and possibly several more) in which the

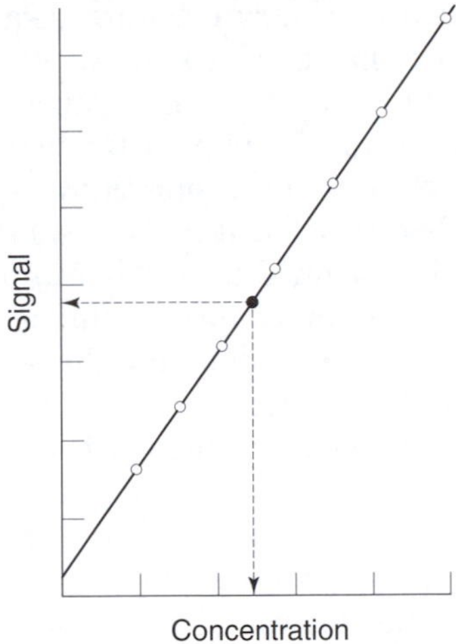

Figure 5.1 Calibration procedure in instrumental analysis: ○ calibration points; ● test sample.

concentration of the analyte is *known*. These calibration standards are measured in the analytical instrument under the same conditions as those subsequently used for the test (i.e. the 'unknown') materials. Once the calibration graph has been established the analyte concentration in any test material can be obtained, as shown in Figure 5.1, by interpolation. This general procedure raises several important statistical questions:

1. Is the calibration graph linear? If it is a curve, what is the form of the curve?
2. Bearing in mind that each of the points on the calibration graph is subject to errors, what is the best straight line (or curve) through these points?
3. Assuming that the calibration plot is actually linear, what are the errors and confidence limits for the slope and the intercept of the line?
4. When the calibration plot is used for the analysis of a test material, what are the errors and confidence limits for the determined concentration?
5. What is the *limit of detection* of the method? That is, what is the least concentration of the analyte that can be detected with a predetermined level of confidence?

Before tackling these questions in detail, we must consider a number of aspects of plotting calibration graphs. Firstly, it is usually essential that the calibration standards cover the whole range of concentrations required in the subsequent analyses. With the important exception of the 'method of standard additions', which is treated separately in a later section, concentrations of test materials are normally determined by interpolation and *not* by extrapolation. Secondly, it is crucially important to include the value for a 'blank' in the calibration curve. The blank contains no deliberately added analyte, but does contain the same solvent, reagents, etc., as the other test materials, and is subjected to exactly the same sequence of analytical procedures. The instrument signal given by the blank will sometimes not be zero. This signal is subject to errors like all the other points on the calibration plot, so it is wrong in principle to subtract the blank value from the other standard values before plotting the calibration graph. This is because, as shown in Chapter 2, when two quantities are subtracted, the error in the final result cannot also be obtained by simple subtraction. Subtracting the blank value from each of the other instrument signals before plotting the graph thus gives incorrect information on the errors in the calibration process. Finally, it should be noted that the calibration curve is always plotted with the instrument signals on the vertical (y) axis and the standard concentrations on the horizontal (x) axis. This is because many of the procedures to be described in the following sections *assume* that all the errors are in the y-values and that the standard concentrations (x-values) are error-free. In many routine instrumental analyses this assumption may well be justified. The standards can be made up with an error of ca. 0.1% or better (see Chapter 1), whereas the instrumental measurements themselves might have a coefficient of variation of 2–3% or worse. So the x-axis error is indeed negligible compared with that of the y-axis. In recent years, however, the advent of high-precision automatic methods with coefficients of variation of 0.5% or better has put the assumption under question, and has led some users to make up their standard solutions by weight rather than by the less accurate combination of weight and volume. This approach is intended to ensure that the x-axis errors remain small compared with those of the y-axis.

Other assumptions usually made are that (a) if several measurements are made on a standard material, the resulting y-values have a normal or Gaussian error distribution; and (b) that the magnitude of the errors in the y-values is independent of the analyte concentration. The first of these two assumptions is usually sound, but the second requires further discussion. If true, it implies that all the points on the graph should have equal *weight* in our calculations, i.e. that it is equally important for the line to pass close to points with high y-values and to those with low y-values. Such calibration graphs are said to be *unweighted*, and are treated in Sections 5.4–5.8. However in practice the y-value errors often increase as the analyte concentration increases. This means that the calibration points should have unequal weight in our calculation, as it is more important for the line to pass close to the points where the errors are least. These *weighted* calculations are now becoming rather more common despite their additional complexity, and are treated in Section 5.10.

In subsequent sections we shall assume that straight line calibration graphs take the algebraic form:

$$y = a + bx \tag{5.1}$$

where b is the slope of the line and a its intercept on the y-axis. The individual points on the line will be referred to as (x_1, y_1 – normally the 'blank' reading), (x_2, y_2), $(x_3, y_3) \ldots (x_i, y_i) \ldots (x_n, y_n)$, i.e. there are n points altogether. The mean of the x-values is, as usual, called $\bar{x}$, and the mean of the y-values is $\bar{y}$: the position $(\bar{x}, \bar{y})$ is then known as the 'centroid' of all the points.

5.3 The product–moment correlation coefficient

In this section we discuss the first problem listed in the previous section – is the calibration plot linear? A common method of estimating how well the experimental points fit a straight line is to calculate the **product–moment correlation coefficient**, r. This statistic is often referred to simply as the 'correlation coefficient' because in quantitative sciences it is by far the most commonly used type of correlation coefficient. We shall, however, meet other types of correlation coefficient in Chapter 6. The value of r is given by:

Product–moment correlation coefficient,

$$r = \frac{\sum_i \{(x_i - \bar{x})(y_i - \bar{y})\}}{\left\{\left[\sum_i (x_i - \bar{x})^2\right]\left[\sum_i (y_i - \bar{y})^2\right]\right\}^{1/2}} \tag{5.2}$$

It can be shown that r can take values in the range $-1 \leq r \leq +1$. As indicated in Figure 5.2 an r-value of -1 describes perfect negative correlation, i.e. all the experimental points lie on a straight line of negative slope. Similarly, when $r = +1$ we have perfect positive correlation, all the points lying exactly on a straight line of positive slope. When there is no correlation between x and y the value of r is close to zero. In analytical practice, calibration graphs frequently give numerical r-values greater than 0.99, and r-values less than about 0.90 are relatively uncommon. A typical example of a calculation of r illustrates a number of important points.

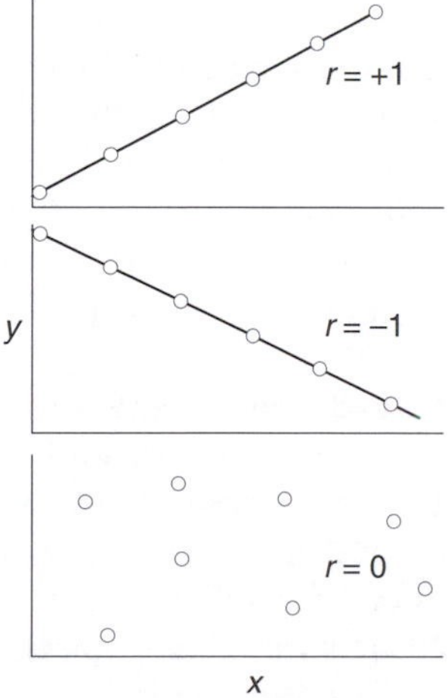

Figure 5.2 The product–moment correlation coefficient, r.

EXAMPLE 5.3.1

Standard aqueous solutions of fluorescein are examined in a fluorescence spectrometer, and yield the following fluorescence intensifies (in arbitrary units):

Fluorescence intensities:	2.1	5.0	9.0	12.6	17.3	21.0	24.7
Concentration, pg ml^{-1}	0	2	4	6	8	10	12

Determine the correlation coefficient, r.

In practice, such calculations will almost certainly be performed on a calculator or computer, alongside other calculations covered below, but it is important and instructive to examine a manually calculated result. The data are presented in a table, as follows:

	x_i	y_i	$x_i - \bar{x}$	$(x_i - \bar{x})^2$	$y_i - \bar{y}$	$(y_i - \bar{y})^2$	$(x_i - \bar{x})(y_i - \bar{y})$
	0	2.1	−6	36	−11.0	121.00	66.0
	2	5.0	−4	16	−8.1	65.61	32.4
	4	9.0	−2	4	−4.1	16.81	8.2
	6	12.6	0	0	−0.5	0.25	0
	8	17.3	2	4	4.2	17.64	8.4
	10	21.0	4	16	7.9	62.41	31.6
	12	24.7	6	36	11.6	134.56	69.6
Sums:	42	91.7	0	112	0	418.28	216.2

The figures below the line at the foot of the columns are in each case the sums of the figures in the table: note that $\sum(x_i - \bar{x})$ and $\sum(y_i - \bar{y})$ are both zero. Using these totals in conjunction with equation (5.2), we have:

$$r = \frac{216.2}{\sqrt{112 \times 418.28}} = \frac{216.2}{216.44} = 0.9989$$

Two observations follow from this example. Figure 5.3 shows that, although several of the points deviate noticeably from the 'best' straight line (calculated using the principles of the following section) the r-value is very close to 1. Experience shows that even quite poor-looking calibration plots give very high r-values. In such cases the numerator and denominator in equation (5.2) are nearly equal. It is thus very important to perform the calculation with an adequate number of significant figures. In the example given neglect of the figures after the decimal point would have given an obviously incorrect r-value of exactly 1, and the use of only one place of decimals would have given the incorrect r-value of 0.9991. This point is especially important when a calculator or computer is used to determine r: it is necessary to ensure that such devices provide sufficient figures.

Although correlation coefficients are simple to calculate, they are all too easily misinterpreted. It must always be borne in mind that the use of equation (5.2) will generate an r-value even if the data are patently non-linear in character. Figure 5.4

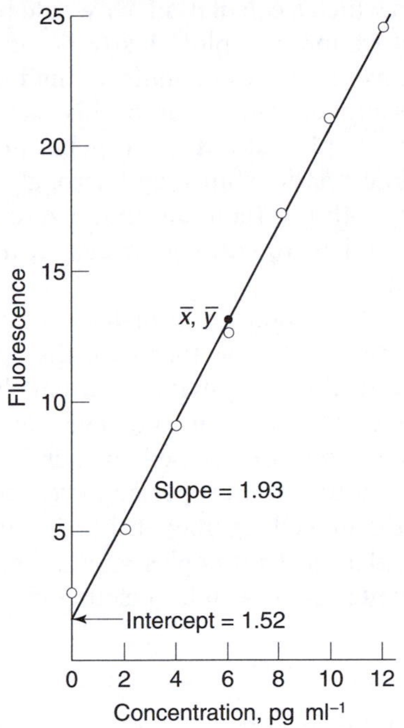

Figure 5.3 Calibration plot for the data in Example 5.3.1.

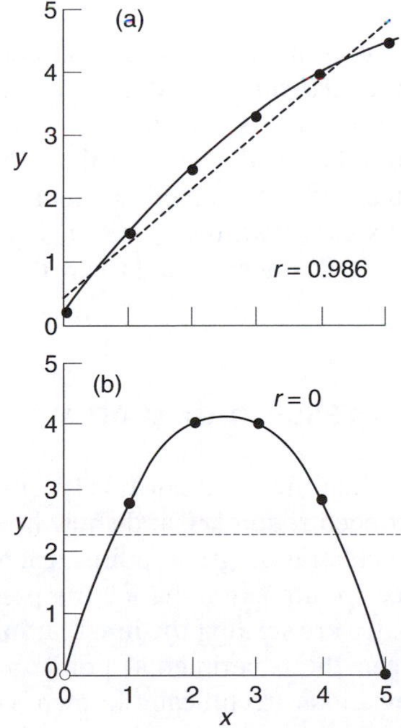

Figure 5.4 Misinterpretation of the correlation coefficient, *r*.

shows two examples in which a calculation of r would be misleading. In Figure 5.4a, the points of the calibration plot clearly lie on a curve; this curve is sufficiently gentle, however, to yield quite a high correlation coefficient when equation (5.2) is applied. The lesson of this example is that the calibration curve must *always* be plotted (on graph paper or a computer monitor): otherwise a straight-line relationship might wrongly be deduced from the calculation of r. Figure 5.4b is a reminder that a zero correlation coefficient does not mean that y and x are entirely unrelated; it only means that they are not *linearly* related.

As we have seen, r-values obtained in instrumental analysis are normally very high, so a calculated value, together with the calibration plot itself, is often sufficient to assure the analyst that a useful linear relationship has been obtained. In some circumstances, however, much lower r-values are obtained: one such situation is discussed further in Section 5.9. In these cases it will be necessary to use a proper statistical test to see whether the correlation coefficient is indeed significant, bearing in mind the number of points used in the calculation. The simplest method of doing this is to calculate a t-value (see Chapter 3 for a fuller discussion of the t-test), using the following equation:

To test for a significant correlation, i.e. H_0 = zero correlation, calculate

$$t = \frac{|r|\sqrt{n-2}}{\sqrt{1-r^2}} \tag{5.3}$$

The calculated value of t is compared with the tabulated value at the desired significance level, using a *two-sided* t-test and $(n - 2)$ degrees of freedom. The null hypothesis in this case is that there is no correlation between x and y. If the calculated value of t is greater than the tabulated value, the null hypothesis is rejected and we conclude in such a case that a significant correlation does exist. As expected, the closer $|r|$ is to 1, i.e. as the straight line relationship becomes stronger, the larger the values of t that are obtained.

5.4 The line of regression of y on x

In this section we assume that there is a linear relationship between the analytical signal (y) and the concentration (x), and show how to calculate the 'best' straight line through the calibration graph points, each of which is subject to experimental error. Since we are assuming for the present that all the errors are in y (cf. Section 5.2), we are seeking the line that minimizes the deviations in the y-direction between the experimental points and the calculated line. Since some of these deviations (technically known as the *y-residuals* – see below) will be positive and some negative, it is sensible to seek to minimize the **sum of the squares of the residuals**, since these squares will all be positive. This explains the frequent use of the term **method of least squares** for the procedure. The straight line required is calculated on this principle: as a result it is found that the line must pass through the centroid of the points $(\bar{x}, \bar{y})$.

It can be shown that the least squares straight line is given by:

$$\text{Slope of least squares line:} \quad b = \frac{\sum_i \{(x_i - \bar{x})(y_i - \bar{y})\}}{\sum_i (x_i - \bar{x})^2} \tag{5.4}$$

$$\text{Intercept of least squares line:} \quad a = \bar{y} - b\bar{x} \tag{5.5}$$

Notice that equation (5.4) contains some of the terms used in equation (5.2), previously used to calculate r: this facilitates calculator or computer operations. The line determined from equations (5.4) and (5.5) is known as **the line of regression of *y* on *x***, i.e. the line indicating how y varies when x is set to chosen values. It is very important to notice that the line of regression of x on y *is not the same line* (except in the highly improbable case where all the points lie exactly on a straight line, when $r = 1$ exactly). The line of regression of x on y (which also passes through the centroid of the points) assumes that all the errors occur in the x-direction. If we maintain rigidly the convention that the analytical signal is always plotted on the y-axis and the concentration on the x-axis, it is always the line of regression of y on x that we must use in calibration experiments.

EXAMPLE 5.4.1

Calculate the slope and intercept of the regression line for the data given in the previous example (see Section 5.3).

In Section 5.3 we calculated that, for this calibration curve:

$$\sum_i (x_i - \bar{x})(y_i - \bar{y}) = 216.2; \quad \sum_i (x_i - \bar{x})^2 = 112; \quad \bar{x} = 6; \quad y = 13.1$$

Using equations (5.4) and (5.5) we calculate that

$$b = 216.2/112 = 1.93$$

$$a = 13.1 - (1.93 \times 6) = 13.1 - 11.58 = 1.52$$

The equation for the regression line is thus $y = 1.93x + 1.52$.

The results of the slope and intercept calculations are depicted in Figure 5.3. Again it is important to emphasize that equations (5.4) and (5.5) must not be misused – they will only give useful results when prior study (calculation of r and a visual inspection of the points) has indicated that a straight line relationship is realistic for the experiment in question.

Non-parametric methods (i.e. methods that make no assumptions about the nature of the error distribution) can also be used to calculate regression lines, and this topic is treated in Chapter 6.

5.5 Errors in the slope and intercept of the regression line

The line of regression calculated in the previous section will in practice be used to estimate the concentrations of test materials by interpolation, and perhaps also to estimate the limit of detection of the analytical procedure. The random errors in the values for the slope and intercept are thus of importance, and the equations used to calculate them are now considered. We must first calculate the statistic $s_{y/x}$, which estimates the random errors in the y-direction.

$$s_{y/x} = \sqrt{\frac{\sum_i (y_i - \hat{y}_i)^2}{n-2}} \tag{5.6}$$

It will be seen that this equation utilizes the y-residuals, $y_i - \hat{y}_i$, where the $\hat{y}_i$-values are the points on the calculated regression line corresponding to the individual x-values, i.e. the 'fitted' y-values (Figure 5.5). The $\hat{y}_i$-value for a given value of x is of course readily calculated from the regression equation. Equation (5.6) is clearly similar in form to the equation for the standard deviation of a set of repeated measurements [equation (2.2)]. The former differs in that deviations, $(y_i - \bar{y})$, are replaced by residuals, $y_i - \hat{y}_i$, and the denominator contains the term $(n - 2)$ rather than $(n - 1)$. In linear regression calculations the number of degrees of freedom (see Section 2.7) is $(n - 2)$. This reflects the obvious consideration that only one straight line can be drawn through two points.

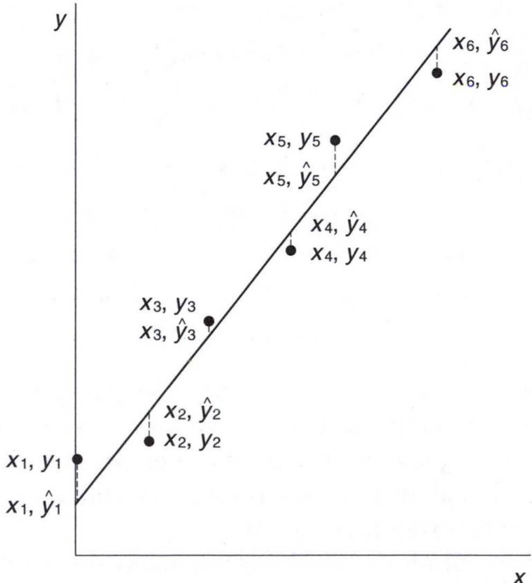

Figure 5.5 The y-residuals of a regression line.

Armed with a value for $s_{y/x}$ we can now calculate s_b and s_a, the standard deviations for the slope (b) and the intercept (a). These are given by:

$$\text{Standard deviation of slope:} \quad s_b = \frac{s_{y/x}}{\sqrt{\sum_i (x_i - \bar{x})^2}} \tag{5.7}$$

$$\text{Standard deviation of intercept:} \quad s_a = s_{y/x}\sqrt{\frac{\sum_i x_i^2}{n\sum_i (x_i - \bar{x})^2}} \tag{5.8}$$

Note again that the term $\sum_i (x_i - \bar{x})^2$ appears in both these equations. The values of s_b and s_a can be used in the usual way (see Chapter 2) to estimate confidence limits for the slope and intercept. Thus the confidence limits for the slope of the line are given by $b \pm t_{(n-2)}s_b$, where the t-value is taken at the desired confidence level and $(n - 2)$ degrees of freedom. Similarly the confidence limits for the intercept are given by $a \pm t_{(n-2)}s_a$.

EXAMPLE 5.5.1

Calculate the standard deviations and confidence limits of the slope and intercept of the regression line calculated in Section 5.4.

This calculation may not be accessible on a simple calculator, but suitable computer software is available. Here we perform the calculation manually, using a tabular layout.

x_i	x_i^2	y_i	$\hat{y}_i$	$\lvert y_i - \hat{y}_i \rvert$	$(y_i - \hat{y}_i)^2$
0	0	2.1	1.52	0.58	0.3364
2	4	5.0	5.38	0.38	0.1444
4	16	9.0	9.24	0.24	0.0576
6	36	12.6	13.10	0.50	0.2500
8	64	17.3	16.96	0.34	0.1156
10	100	21.0	20.82	0.18	0.0324
12	144	24.7	24.68	0.02	0.0004

$$\sum_i x_i^2 = 364 \qquad \sum_i (y_i - \hat{y}_i)^2 = 0.9368$$

From the table and using equation (5.6) we obtain

$$s_{y/x} = \sqrt{0.9368/5} = \sqrt{0.18736} = 0.4329$$

From Section 5.3 we know that $\sum_i (x_i - \bar{x})^2 = 112$, and equation (5.7) can be used to show that

$$s_b = 0.4329/\sqrt{112} = 0.4329/10.58 = 0.0409$$

The t-value for $(n-2)=5$ degrees of freedom and the 95% confidence level is 2.57 (Table A.2). The 95% confidence limits for b are thus:

$$b = 1.93 \pm (2.57 \times 0.0409) = 1.93 \pm 0.11$$

Equation (5.8) requires knowledge of $\sum_i x_i^2$, calculated as 364 from the table. We can thus write:

$$s_a = 0.4329\sqrt{\frac{364}{7 \times 112}} = 0.2950$$

so the 95% confidence limits are:

$$a = 1.52 \pm (2.57 \times 0.2950) = 1.52 \pm 0.76$$

In this example, the number of significant figures necessary was not large, but it is always a useful precaution to use the maximum available number of significant figures during such a calculation, rounding only at the end.

There is no necessity in practice for the manual calculation of all these results, which would clearly be too tedious for routine use. The application of a spreadsheet program to some regression data is demonstrated in Section 5.9. Every advantage should also be taken of the extra facilities provided by programs such as Minitab, for example plots of residuals against x or $\hat{y}$ values, normal probability plots for the residuals, etc. (see also Section 5.15).

5.6 Calculation of a concentration and its random error

Once the slope and intercept of the regression line have been determined, it is very simple to calculate the concentration (x-value) corresponding to any measured instrument signal (y-value). But it will also be necessary to find the error associated with this concentration estimate. Calculation of the x-value from the given y-value using equation (5.1) involves the use of both the slope (b) and the intercept (a) and, as we saw in the previous section, both these values are subject to error. Moreover, the instrument signal derived from any test material is also subject to random errors. As a result, the determination of the overall error in the corresponding concentration is extremely complex, and most workers use the following approximate formula:

$$s_{x_0} = \frac{s_{y/x}}{b}\sqrt{1 + \frac{1}{n} + \frac{(y_0 - \bar{y})^2}{b^2 \sum_i (x_i - \bar{x})^2}} \qquad (5.9)$$

In this equation, y_0 is the experimental value of y from which the concentration value x_0 is to be determined, s_{x_0} is the estimated standard deviation of x_0, and the other symbols have their usual meanings. In some cases an analyst may make several readings to obtain the value of y_0: if there are m such readings, then the equation for s_{x_0} becomes:

$$s_{x_0} = \frac{s_{y/x}}{b}\sqrt{\frac{1}{m} + \frac{1}{n} + \frac{(y_0 - \bar{y})^2}{b^2 \sum_i (x_i - \bar{x})^2}} \qquad (5.10)$$

As expected, equation (5.10) reduces to equation (5.9) if $m = 1$. As always, confidence limits can be calculated as $x_0 \pm t_{(n-2)} s_{x_0}$, with $(n - 2)$ degrees of freedom. Again, a simple computer program will perform all these calculations, but most calculators will not be adequate.

EXAMPLE 5.6.1

Using the data from the Section 5.3 example, determine x_0 and s_{x_0} values and x_0 confidence limits for solutions with fluorescence intensities of 2.9, 13.5 and 23.0 units.

The x_0 values are easily calculated by using the regression equation determined in Section 5.4, $y = 1.93x + 1.52$. Substituting the y_0-values 2.9. 13.5 and 23.0, we obtain x_0-values of 0.72, 6.21 and 11.13 pg ml^{-1} respectively.

To obtain the s_{x_0}-values corresponding to these x_0-values we use equation (5.9), recalling from the preceding sections that $n = 7$, $b = 1.93$, $s_{y/x} = 0.4329$, $\bar{y} = 13.1$, and $\sum_i (x_i - \bar{x})^2 = 112$. The y_0 values 2.9, 13.5 and 23.0 then yield s_{x_0}-values of 0.26, 0.24 and 0.26 respectively. The corresponding 95% confidence limits ($t_5 = 2.57$) are 0.72 ± 0.68, 6.21 ± 0.62, and 11.13 ± 0.68 pg ml^{-1} respectively.

This example illustrates an important point. It shows that the confidence limits are rather smaller (i.e. better) for the result $y_0 = 13.5$ than for the other two y_0-values. Inspection of equation (5.9) confirms that as y_0 approaches $\bar{y}$, the third term inside the bracket approaches zero, and s_{x_0} thus approaches a minimum value. The general form of the confidence limits for a calculated concentration is shown in Figure 5.6. Thus in practice a calibration experiment of this type will give the most precise results when the measured instrument signal corresponds to a point close to the centroid of the regression line.

If we wish to improve (i.e. narrow) the confidence limits in this calibration experiment, equations (5.9) and (5.10) show that at least two approaches should be considered. We could increase n, the number of calibration points on the regression line, and/or we could make more than one measurement of y_0, using the mean value of m such measurements in the calculation of x_0. The results of such procedures can be assessed by considering the three terms inside the brackets in the two equations. In the example above, the dominant term in all three calculations is the first one – unity. It follows

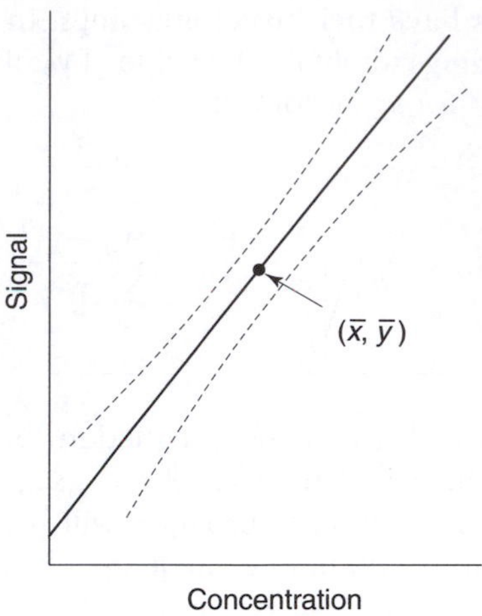

Figure 5.6 General form of the confidence limits for a concentration determined by using an unweighted regression line.

that in this case (and many others) an improvement in precision might be made by measuring y_0 several times and using equation (5.10) rather than equation (5.9). If, for example, the y_0-value of 13.5 had been calculated as the mean of four determinations, then the s_{x_0}-value and the confidence limits would have been 0.14 and 6.21 ± 0.36 respectively, both results indicating substantially improved precision. Of course, making too many replicate measurements (assuming that sufficient sample is available) generates much more work for only a small additional benefit: the reader should verify that eight measurements of y_0 would produce an s_{x_0}-value of 0.12 and confidence limits of 6.21 ± 0.30.

The effect of n, the number of calibration points, on the confidence limits of the concentration determination is more complex. This is because we also have to take into account accompanying changes in the value of t. Use of a large number of calibration samples involves the task of preparing many accurate standards for only marginally increased precision (cf. the effects of increasing m, described in the previous paragraph). On the other hand, small values of n are not permissible. In such cases $1/n$ will be larger and the number of degrees of freedom, $(n - 2)$, will become very small, necessitating the use of very large t-values in the calculation of the confidence limits. In many experiments, as in the example given, six or so calibration points will be adequate, the analyst gaining extra precision if necessary by repeated measurements of y_0.

5.7 Limits of detection

As we have seen, one of the principal benefits of using instrumental methods of analysis is that they are capable of detecting and determining trace and ultra-trace quantities of analytes. These benefits have led to the appreciation of the importance of very low concentrations of many materials, for example in biological and environmental samples, and thus to the development of many further techniques in which lower limits of detection are a major criterion of successful application. It is therefore evident that statistical methods for assessing and comparing limits of detection are of importance. In general terms, the limit of detection of an analyte may be described as that concentration which gives an instrument signal (y) *significantly different* from the 'blank' or 'background' signal. This description gives the analyst a good deal of freedom to decide the exact definition of the limit of detection, based on a suitable interpretation of the phrase 'significantly different'. There is still no full agreement between researchers, publishers, and professional and statutory bodies on this point. But there is an increasing trend to define the limit of detection as the analyte concentration giving a signal equal to the blank signal, y_B, plus three standard deviations of the blank, s_B:

$$\text{Limit of detection} = y_B + 3s_B \tag{5.11}$$

The significance of this last definition is illustrated in more detail in Figure 5.7. An analyst studying trace concentrations is confronted with two problems: it is important to avoid claiming the presence of the analyte when it is actually absent, but it is equally important to avoid reporting that the analyte is absent when it is in fact present. (The situation is analogous to the occurrence of Type I and Type II errors in significance tests – see Section 3.13.) The possibility of each of these errors must be minimized by a sensible definition of a limit of detection. In the figure, curve A represents the normal distribution of measured values of the blank signal. It would be possible to identify a point, $y = \text{P}$, towards the upper edge of this distribution, and claim that a signal greater than this was unlikely to be due to the blank, whereas a signal less than P would be assumed to indicate a blank sample. However, for a sample giving an average signal P, 50% of the observed signals will be less than this, since the signal will have a normal distribution (of the same shape as that for the blank – see below) extending below P (curve B). The probability of concluding that this sample does not differ from the blank

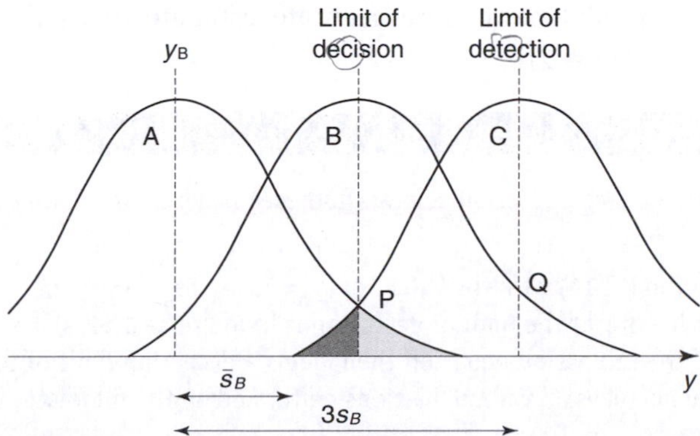

Figure 5.7 Definitions of the limit of decision and the limit of detection.

when in fact it does, is therefore 50%. Point P, which has been called the limit of decision, is thus unsatisfactory as a limit of detection, since it solves the first of the problems mentioned above, but not the second. A more suitable point is at $y = Q$, such that Q is twice as far as P from y_B. It may be shown that if $y_B - Q$ is 3.28 times the standard deviation of the blank, s_B, then the probability of each of the two kinds of error occurring (indicated by the shaded areas in Figure 5.7) is only 5%. If, as suggested by equation (5.11), the distance $y_B - Q$ is only $3s_B$, the probability of each error is about 7%: many analysts would consider that this is a reasonable definition of a limit of detection.

It must be re-emphasized that this definition of a limit of detection is quite arbitrary, and it is entirely open to an analyst to provide an alternative definition for a particular purpose. For example, there may be occasions when an analyst is anxious to avoid at all costs the possibility of reporting the absence of the analyte when it is in fact present, but is relatively unworried about the opposite error. It is clear that whenever a limit of detection is cited in a paper or report, the definition used to obtain it must also be provided. Some attempts have been made to define a further limit, the 'limit of quantitation' (or 'limit of determination'), which is regarded as the lower limit for precise quantitative measurements, as opposed to qualitative detection. A value of $y_B + 10s_B$ has been suggested for this limit, but it is not very widely used.

We must now discuss how the terms y_B and s_B are obtained in practice when a regression line is used for calibration as described in the preceding sections. A fundamental assumption of the unweighted least-squares method is that each point on the plot (including the point representing the blank or background) has a normally distributed variation (in the y-direction only) with a standard deviation estimated by $s_{y/x}$ [equation (5.6)]. This is the justification for drawing the normal distribution curves with the same width in Figure 5.7. It is therefore appropriate to use $s_{y/x}$ in place of s_B in the estimation of the limit of detection. It is, of course, possible to perform the blank experiment several times and obtain an independent value for s_B, and if our underlying assumptions are correct these two methods of estimating s_B should not differ significantly. But multiple determinations of the blank are time-consuming and the use of $s_{y/x}$ is quite suitable in practice. The value of a, the calculated intercept, can be used as an estimate of y_B, the blank signal itself; it should be a more accurate estimate of y_B than the single measured blank value, y_1.

EXAMPLE 5.7.1

Estimate the limit of detection for the fluorescein determination studied in the previous sections.

We use equation (5. 11) with the values of $y_B(= a)$ and $s_B(= s_{y/x})$ previously calculated. The value of y at the limit of detection is found to be $1.52 + 3 \times 0.4329$, i.e. 2.82. Use of the regression equation then yields a detection limit of 0.67 pg ml^{-1}. Figure 5.8 summarizes all the calculations performed on the fluorescein determination data.

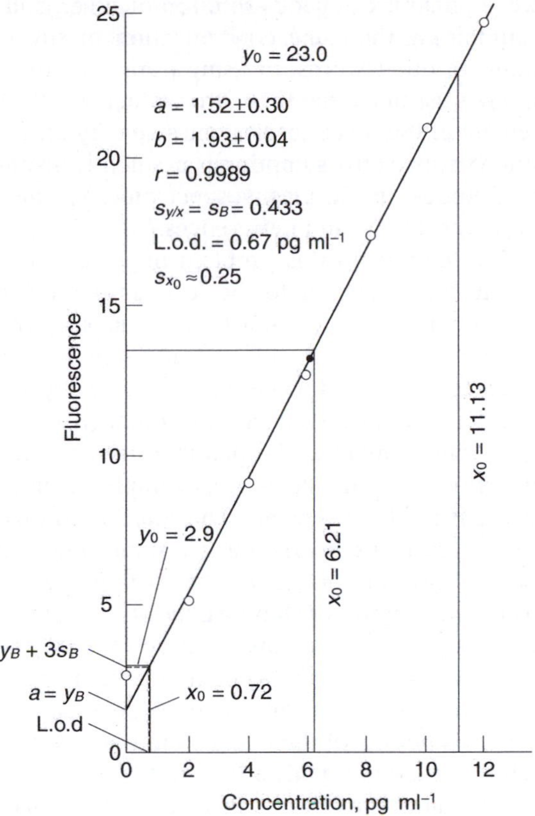

Figure 5.8 Summary of the calculations using the data in Example 5.3.1.

It is important to avoid confusing the limit of detection of a technique with its sensitivity. This very common source of confusion probably arises because there is no single generally accepted English word synonymous with 'having a low limit of detection'. The word 'sensitive' is generally used for this purpose, giving rise to much ambiguity. The sensitivity of a technique is correctly defined as the *slope* of the calibration graph and, provided the plot is linear, can be measured at any point on it. In contrast, the limit of detection of a method is calculated with the aid of the section of the plot close to the origin, and utilizes both the slope and the $s_{y/x}$ value.

5.8 The method of standard additions

Suppose that we wish to determine the concentration of silver in samples of photographic waste by atomic-absorption spectrometry. Using the methods of the previous sections, an analyst could calibrate the spectrometer with some aqueous solutions of a pure silver salt and use the resulting calibration graph in the determination of the silver in the test samples. This method is only valid, however, if a pure aqueous solution of silver, and a photographic waste sample containing the same concentration of silver, give the same absorbance values. In other words, in using pure solutions to establish the calibration graph it is assumed that there are no 'matrix effects', i.e. no reduction or enhancement of the silver absorbance signal by other components. In many areas of analysis such an assumption is frequently invalid. Matrix effects occur even with methods such as plasma spectrometry, which have a reputation for being relatively free from interferences.

The first possible solution to this problem might be to take a sample of photographic waste that is similar to the test sample, but free from silver, and add known amounts of a silver salt to it to make up the standard solutions. The calibration graph will then be set up using an apparently suitable matrix. In many cases, however, this *matrix matching* approach is impracticable. It will not eliminate matrix effects that differ in magnitude from one sample to another, and it may not be possible even to obtain a sample of the matrix that contains no analyte – for example, a silver-free sample of photographic waste is unlikely to occur! The solution to this problem is that all the analytical measurements, including the establishment of the calibration graph, must in some way be performed *using the sample itself*. This is achieved in practice by using the **method of standard additions**. The method is widely practised in atomic absorption and emission spectrometry and has also found application in electrochemical analysis and many other areas. Equal volumes of the sample solution are taken, all but one are separately 'spiked' with known and different amounts of the analyte, and *all* are then diluted to the same volume. The instrument signals are then determined for all these solutions and the results plotted as shown in Figure 5.9. As usual, the signal is plotted on the y-axis; in this case the x-axis is graduated in terms of the amounts of analyte *added* (either as an absolute weight or as a concentration). The (unweighted) regression line is calculated in the normal way, but space is provided for it to be extrapolated to the point on the x-axis at which $y = 0$. This negative intercept on the x-axis corresponds to the amount of the analyte in the test sample. Inspection of the figure shows that this value is

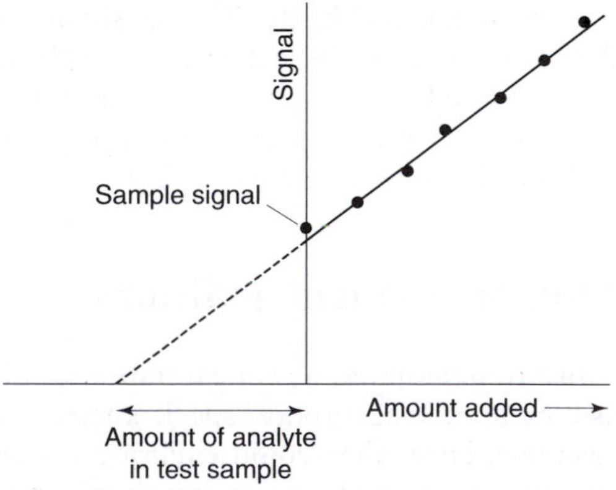

Figure 5.9 The method of standard additions.

given by a/b, the ratio of the intercept and the slope of the regression line. Since both a and b are subject to error (Section 5.5) the calculated concentration is clearly subject to error as well. In this case, however, the amount is not predicted from a single measured value of y, so the formula for the standard deviation, s_{x_E}, of the extrapolated x-value (x_E) is not the same as that in equation (5.9).

$$s_{x_E} = \frac{s_{y/x}}{b}\sqrt{\frac{1}{n} + \frac{\bar{y}^2}{b^2 \sum_i (x_i - \bar{x})^2}} \qquad (5.12)$$

Increasing the value of n again improves the precision of the estimated concentration: in general at least six points should be used in a standard-additions experiment. Moreover, the precision is improved by maximizing $\sum_i (x_i - \bar{x})^2$, so the calibration solutions should, if possible, cover a considerable range. Confidence limits for x_E can as before be determined as $x_E \pm t_{(n-2)} s_{x_E}$.

EXAMPLE 5.8.1

The silver concentration in a sample of photographic waste was determined by atomic-absorption spectrometry with the method of standard additions. The following results were obtained.

Added Ag: µg added per ml of *original* sample solution	0	5	10	15	20	25	30
Absorbance	0.32	0.41	0.52	0.60	0.70	0.77	0.89

Determine the concentration of silver in the sample, and obtain 95% confidence limits for this concentration.

Equations (5.4) and (5.5) yield $a = 0.3218$ and $b = 0.0186$. The ratio of these figures gives the silver concentration in the test sample as 17.3 µg ml^{-1}. The confidence limits for this result can be determined with the aid of equation (5.12). Here $s_{y/x}$ is 0.01094, $\bar{y} = 0.6014$, and $\sum_i (x_i - \bar{x})^2 = 700$. The value of s_{x_E} is thus 0.749 and the confidence limits are $17.3 \pm 2.57 \times 0.749$, i.e. 17.3 ± 1.9 µg ml^{-1}.

Although it is an elegant approach to the common problem of matrix interference effects, the method of standard additions has a number of disadvantages. The principal one is that each test sample requires its own calibration graph, in contrast to conventional calibration experiments, where one graph can provide concentration values for many test samples. The standard additions method may also use larger quantities of sample than other methods. In statistical terms it is an extrapolation method, and in principle less precise than interpolation techniques. In practice, the loss of precision is not very serious.

5.9 Use of regression lines for comparing analytical methods

If an analytical chemist develops a new method for the determination of a particular analyte, the method must be validated by (amongst other techniques) applying it to a series of materials already studied using another reputable or standard procedure. The main aim of such a comparison will be the identification of systematic errors - does the new method give results that are significantly higher or lower than the established procedure? In cases where an analysis is repeated several times over a very limited concentration range, such a comparison can be made using the statistical tests described in Sections 3.3 and 3.4. Such procedures will not be appropriate in instrumental analyses, which are often used over large concentration ranges.

When two methods are to be compared at different analyte concentrations the procedure illustrated in Figure 5.10 is normally adopted. One axis of a regression graph is used for the results obtained by the new method, and the other axis for the results obtained by applying the reference or comparison method to the same samples. (The question of which axis should be allocated to each method is discussed further below.) Each point on the graph thus represents a single sample analysed by two separate methods. (Sometimes each method is applied just once to each test sample, while in other cases

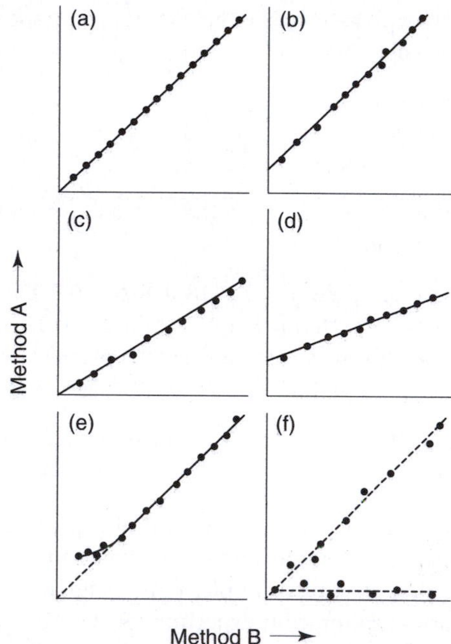

Figure 5.10 Use of a regression line to compare two analytical methods: (a) shows perfect agreement between the two methods for all the samples; (b)–(f) illustrate the results of various types of systematic error (see text).

replicate measurements are used in the comparisons.) The methods of the preceding sections are then applied to calculate the slope (b), the intercept (a) and the product–moment correlation coefficient (r) of the regression line. It is clear that if each sample yields an identical result with both analytical methods the regression line will have a zero intercept, and a slope and a correlation coefficient of 1 (Figure 5.10a). In practice, of course, this never occurs: even if systematic errors are entirely absent, random errors ensure that the two analytical procedures will not give results in exact agreement for all the samples.

Deviations from the 'ideal' situation ($a = 0$, $b = r = 1$) can occur in a number of different ways. Firstly, it is possible that the regression line will have a slope of 1, but a non-zero intercept. That is, one method of analysis may yield a result higher or lower than the other by a fixed amount. Such an error might occur if the background signal for one of the methods was wrongly calculated (Figure 5.10b). A second possibility is that the slope of the regression line is >1 or <1, indicating that a systematic error may be occurring in the slope of one of the individual calibration plots (Figure 5.10c). These two errors may occur simultaneously (Figure 5.10d). Further possible types of systematic error are revealed if the plot is curved (Figure 5.10e). Speciation problems may give surprising results (Figure 5.10f). This type of plot might arise if an analyte occurred in two chemically distinct forms, the proportions of which varied from sample to sample. One of the methods under study (here plotted on the y-axis) might detect only one form of the analyte, while the second method detected both forms.

In practice, the analyst most commonly wishes to test for an intercept differing significantly from zero, and a slope differing significantly from 1. Such tests are performed by determining the confidence limits for a and b, generally at the 95% significance level. The calculation is very similar to that described in Section 5.5, and is most simply performed by using a program such as Excel. This spreadsheet is applied to the following example.

EXAMPLE 5.9.1

The level of phytic acid in 20 urine samples was determined by a new catalytic fluorimetric (CF) method, and the results were compared with those obtained using an established extraction photometric (EP) technique. The following data were obtained (all the results, in mg l^{-1}, are means of triplicate measurements).

(March, J. G., Simonet, B. M. and Grases, F. 1999. Analyst 124: 897–900)

Sample number	*CF result*	*EP result*
1	1.87	1.98
2	2.20	2.31
3	3.15	3.29
4	3.42	3.56
5	1.10	1.23

Sample number	*CF result*	*EP result*
6	1.41	1.57
7	1.84	2.05
8	0.68	0.66
9	0.27	0.31
10	2.80	2.82
11	0.14	0.13
12	3.20	3.15
13	2.70	2.72
14	2.43	2.31
15	1.78	1.92
16	1.53	1.56
17	0.84	0.94
18	2.21	2.27
19	3.10	3.17
20	2.34	2.36

This set of data shows why it is inappropriate to use the paired *t*-test, which evaluates the differences between the pairs of results, in such cases (Section 3.4). The range of phytic acid concentrations (ca. 0.14–3.50 mg l^{-1}) in the urine samples is so large that a fixed discrepancy between the two methods will be of varying significance at different concentrations. Thus a difference between the two techniques of 0.05 mg l^{-1} would not be of great concern at a level of ca. 3.50 mg l^{-1}, but would be more disturbing at the lower end of the concentration range.

Table 5.1 shows the summary output of the Excel spreadsheet used to calculate the regression line for the above data. The CF data have been plotted on the *y*-axis, and the EP results on the *x*-axis (see below). The output shows that the *r*-value (called 'Multiple R' by this program because of its potential application to multiple regression methods) is 0.9967. The intercept is −0.0456, with upper and lower confidence limits of −0.1352 and +0.0440: this range includes the ideal value of zero. The slope of the graph, called 'X variable 1' because *b* is the coefficient of the *x*-term in equation (5.1), is 0.9879, with a 95% confidence interval of 0.9480–1.0279: again this range includes the model value, in this case 1.0. (The remaining output data are not needed in this example, and are discussed further in Section 5.11.) Figure 5.11 shows the regression line with the characteristics summarized above.

Two further points may be mentioned in connection with this example. Firstly, the literature of analytical chemistry shows that authors frequently place great stress on the value of the correlation coefficient in such comparative studies. In the above example, however, it played no direct role in establishing whether or not systematic errors had occurred. Even if the regression line had been slightly curved, the correlation coefficient might still have been

Table 5.1 Excel output for Example 5.9.1

Regression statistics

Multiple R	0.9967
R square	0.9934
Adjusted R square	0.9930
Standard error	0.0825
Observations	20

ANOVA

	df	*SS*	*MS*	*F*	*Significance F*
Regression	1	18.342	18.342	2695.977	4.61926E-21
Residual	18	0.122	0.007		
Total	19	18.465			

	Coefficients	Standard error	t stat	P-value
Intercept	-0.0456	0.0426	-1.070	0.299
X variable 1	0.9879	0.0190	51.923	4.62E-21

	Lower 95%	Upper 95%	Lower 95.0%	Upper 95.0%
Intercept	-0.1352	0.0440	-0.1352	0.0440
X variable 1	0.9480	1.0279	0.9480	1.0279

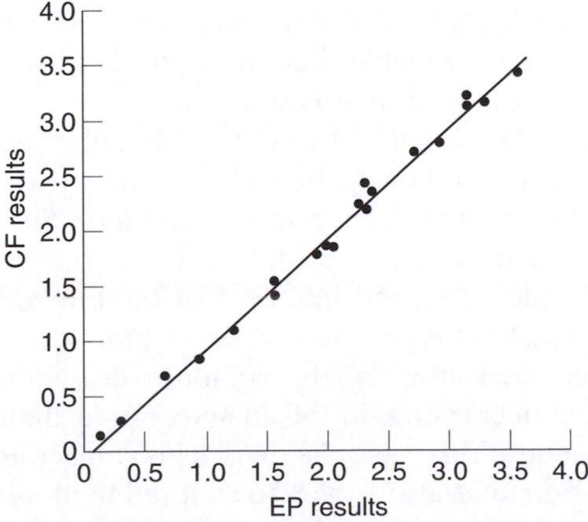

Figure 5.11 Comparison of two analytical methods: data from Example 5.9.1.

close to 1 (see Section 5.3 above). This means that the calculation of r is less important in the present context than the establishment of confidence limits for the slope and the intercept. In some cases it may be found that the r-value is not very close to 1, even though the slope and the intercept are not significantly different from 1 and 0 respectively. Such a result would suggest very poor precision for either one or both of the methods under study. The precisions of the two methods can be determined and compared using the methods of Chapters 2 and 3. In practice it is desirable that this should be done *before* the regression line comparing the methods is plotted – the reason for this is explained below. The second point to note is that it is desirable to compare the methods over the full range of concentrations, as in the example given where the urine samples examined contained phytic acid concentrations that covered the range of interest fairly uniformly.

Although very widely adopted in comparative studies of instrumental methods, the approach described here is open to some theoretical objections. First, as has been emphasized throughout this chapter, the line of regression of y on x is calculated on the assumption that the errors in the x-values are negligible – all errors are assumed to occur in the y-direction. While generally valid in a calibration plot for a single analyte, this assumption is evidently not justified when the regression line is used for comparison purposes: it can be taken as certain that random errors will occur in both analytical methods, i.e. in both the x and y directions. This suggests that the equations used to calculate the regression line itself are not valid. However the regression method is still widely used, as the graphs which result provide valuable information on the nature of any differences between the methods (Figure 5.10). Simulations show, moreover, that the approach does give surprisingly reliable results, provided that the more precise method is plotted on the x-axis (this is why we investigate the precisions of the two methods – see above), and that a reasonable number of points (ca. 10 at least) uniformly covering the concentration range of interest is used. Since the confidence limit calculations are based on $(n - 2)$ degrees of freedom, it is particularly important to avoid small values of n. Methods for plotting regression lines where both x and y are subject to error are available, but in practice are not widely used in comparison studies because of their complexity.

A second objection to using the line of regression of y on x, as calculated in Sections 5.4 and 5.5, in the comparison of two analytical methods is that it also assumes that the error in the y-values is *constant*. Such data are said to be **homoscedastic**. As previously noted, this means that all the points have equal weight when the slope and intercept of the line are calculated. This assumption is obviously likely to be invalid in practice. In many analyses, the data are **heteroscedastic**, i.e. the standard deviation of the y-values increases with the concentration of the analyte, rather than having the same value at all concentrations (see below). This objection to the use of unweighted regression lines also applies to calibration plots for a single analytical procedure. In principle **weighted regression** lines should be used instead, as shown in the next section.

5.10 Weighted regression lines

In this section the application of weighted regression methods is outlined. It is assumed that the weighted regression line is to be used for the determination of a single analyte rather than for the comparison of two separate methods. In any calibration analysis the overall random error of the result will arise from a combination of the error contributions from the several stages of the analysis (see Section 2.11). In some cases this overall error will be dominated by one or more steps in the analysis where the random error is not concentration dependent. In such cases we shall expect the y-direction errors in the calibration curve to be approximately equal for all the points (homoscedasticity), and an unweighted regression calculation is legitimate. In other cases the errors will be approximately proportional to analyte concentration (i.e. the *relative* error will be roughly constant), and in still others (perhaps the commonest situation in practice) the y-direction error will increase as x increases, but less rapidly than the concentration. Both these types of heteroscedastic data should be treated by weighted regression methods. Usually an analyst can only learn from experience whether weighted or unweighted methods are appropriate. Predictions are difficult: examples abound where two apparently similar methods show very different error behaviour. Weighted regression calculations are rather more complex than unweighted ones, and they require more information (or the use of more assumptions). Nonetheless they should be used whenever heteroscedasticity is suspected, and they are now more widely applied than formerly, partly as a result of pressure from regulatory authorities in the pharmaceutical industry and elsewhere.

Figure 5.12 shows the simple situation that arises when the error in a regression calculation is approximately proportional to the concentration of the analyte, i.e. the 'error bars' used to express the random errors at different points on the calibration get larger as the concentration increases. The regression

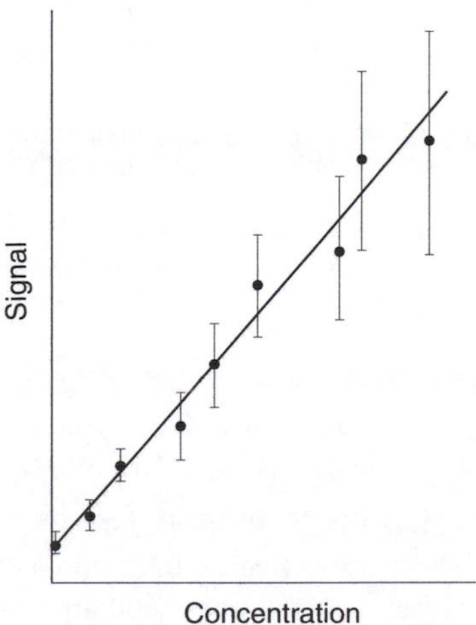

Figure 5.12 The weighting of errors in a regression calculation.

line must be calculated to give additional weight to those points where the error bars are smallest: it is more important for the calculated line to pass close to such points than to pass close to the points representing higher concentrations with the largest errors. This result is achieved by giving each point a weighting inversely proportional to the corresponding variance, s_i^2. (This logical procedure applies to all weighted regression calculations, not just those where the y-direction error is proportional to x.) Thus, if the individual points are denoted by (x_1, y_1), (x_2, y_2), etc. as usual, and the corresponding standard deviations are s_1, s_2, etc., then the individual weights, w_1, w_2, etc., are given by:

$$\text{Weights:} \qquad w_i = \frac{s_i^{-2}}{\sum_i s_i^{-2}/n} \tag{5.13}$$

It will be seen that the weights have been scaled so that their sum is equal to the number of points on the graph: this simplifies the subsequent calculations. The slope and the intercept of the recession line are then given by:

$$\text{Weighted slope:} \qquad b_w = \frac{\sum_i w_i x_i y_i - n\bar{x}_w \bar{y}_w}{\sum_i w_i x_i^2 - n\bar{x}_w^2} \tag{5.14}$$

and

$$\text{Weighted intercept:} \qquad a_w = \bar{y}_w - b\bar{x}_w \tag{5.15}$$

In equation (5.15) $\bar{y}_w$ and $\bar{x}_w$ represent the co-ordinates of the *weighted centroid*, through which the weighted regression line must pass. These co-ordinates are given as expected by $\bar{x}_w = \sum_i w_i x_i / n$ and $\bar{y}_w = \sum_i w_i y_i / n$.

EXAMPLE 5.10.1

Calculate the unweighted and weighted regression lines for the following calibration data. For each line calculate also the concentrations of test samples with absorbances of 0.100 and 0.600.

Concentration, μg ml^{-1}	0	2	4	6	8	10
Standard deviation	0.001	0.004	0.010	0.013	0.017	0.022
Absorbance	0.009	0.158	0.301	0.472	0.577	0.739

Application of equations (5.4) and (5.5) shows that the slope and intercept of the *unweighted* regression line are respectively 0.0725 and 0.0133. The concentrations corresponding to absorbances of 0.100 and 0.600 are then found to be 1.20 and 8.09 μg ml^{-1} respectively.

The *weighted* regression line is a little harder to calculate: in the absence of a suitable computer program it is usual to set up a table as follows.

x_i	y_i	s_i	$1/s_i^2$	w_i	$w_i x_i$	$w_i y_i$	$w_i x_i y_i$	$w_i x_i^2$
0	0.009	0.001	10^6	5.535	0	0.0498	0	0
2	0.158	0.004	62500	0.346	0.692	0.0547	0.1093	1.384
4	0.301	0.010	10000	0.055	0.220	0.0166	0.0662	0.880
6	0.472	0.013	5917	0.033	0.198	0.0156	0.0935	1.188
8	0.577	0.017	3460	0.019	0.152	0.0110	0.0877	1.216
10	0.739	0.022	2066	0.011	0.110	0.0081	0.0813	1.100
Sums			1083943	5.999	1.372	0.1558	0.4380	5.768

These figures give $\bar{y}_w = 0.1558/6 = 0.0260$, and $\bar{x}_w = 1.372/6 = 0.229$. By equation (5.14), b_w is calculated from

$$b_w = \frac{0.438 - (6 \times 0.229 \times 0.026)}{5.768 - [6 \times (0.229)^2]} = 0.0738$$

so a_w is given by $0.0260 - (0.0738 \times 0.229) = 0.0091$.

These values for a_w and b_w can be used to show that absorbance values of 0.100 and 0.600 correspond to concentrations of 1.23 and 8.01 μg ml^{-1} respectively.

Comparison of the results of the unweighted and weighted regression calculations is very instructive. The effects of the weighting process are clear. The weighted centroid $(\bar{x}_w, \bar{y}_w)$ is much closer to the origin of the graph than the unweighted centroid $(\bar{x}, \bar{y})$ and the weighting given to the points nearer the origin (particularly to the first point (0, 0.009) which has the smallest error) ensures that the weighted regression line has an intercept very close to this point. The slope and intercept of the weighted line are remarkably similar to those of the unweighted line, however, with the result that the two methods give very similar values for the concentrations of samples having absorbances of 0.100 and 0.600. It must not be supposed that these similar values arise simply because in this example the experimental points fit a straight line very well. In practice the weighted and unweighted regression lines derived from a set of experimental data have similar slopes and intercepts even if the scatter of the points about the line is substantial.

As a result it might seem that weighted regression calculations have little to recommend them. They require more information (in the form of estimates of the standard deviation at various points on the graph), and are far more complex to execute, but they seem to provide data that are remarkably similar to those obtained from the much simpler unweighted regression method. Such considerations may indeed account for some of the neglect of weighted regression calculations in practice. But an analytical chemist using instrumental methods does not employ regression calculations simply to determine the slope and intercept of the calibration plot and the concentrations of test samples. There is also a need to obtain estimates of the errors or confidence limits of those concentrations, and it is in this context that the weighted regression method provides much more realistic results. In Section 5.6 we used equation (5.9) to estimate the standard deviation (s_{x_0}) and hence the confidence limits of a concentration calculated using a single y-value and an unweighted regression line. Application of this equation to the data in the example above shows that the unweighted confidence limits for the solutions having absorbances of 0.100 and 0.600 are 1.20 ± 0.65 and 8.09 ± 0.63 μg ml^{-1} respectively. As in the example in Section 5.6, these confidence intervals are very similar. In the present example, however, such a result is entirely unrealistic. The experimental data show that the errors of the observed y-values increase as y itself increases, the situation expected for a method having a roughly constant relative standard deviation. We would expect that this increase in s_i with increasing y would also be reflected in the confidence limits of the determined concentrations: the confidence limits for the solution with an absorbance of 0.600 should be much greater (i.e. worse) than those for the solution with an absorbance of 0.100.

In weighted recession calculations, the standard deviation of a predicted concentration is given by:

$$s_{x_{0w}} = \frac{s_{(y/x)w}}{b} \left\{ \frac{1}{w_0} + \frac{1}{n} + \frac{(y_0 - \bar{y}_w)^2}{b^2 \left(\sum_i w_i x_i^2 - n\bar{x}_w^2 \right)} \right\}^{1/2} \qquad (5.16)$$

In this equation, $s_{(y/x)w}$ is given by:

$$s_{(y/x)w} = \left\{ \frac{\sum_i w_i (y_i - \hat{y}_i)^2}{n-2} \right\}^{1/2} \qquad (5.17)$$

and w_0 is a weighting appropriate to the value of y_0. Equations (5.16) and (5.17) are clearly similar in form to equations (5.9) and (5.6). Equation (5.16) confirms that points close to the origin, where the weights are highest, and points near the centroid, where $(y_0 - \bar{y}_w)$ is small, will have the narrowest confidence limits (Figure 5.13). The major difference between equations (5.9) and (5.16) is the term $1/w_0$ in the latter. Since w_0 falls sharply as y increases, this term ensures that the confidence limits increase with increasing y_0, as we expect.

Application of equation (5.16) to the data in the example above shows that the test samples with absorbance of 0.100 and 0.600 have confidence limits for the calculated concentrations of 1.23 ± 0.12 and 8.01 ± 0.72 μg ml^{-1} respectively. The widths of these confidence intervals are proportional to the observed absorbances of the two solutions. In addition the confidence

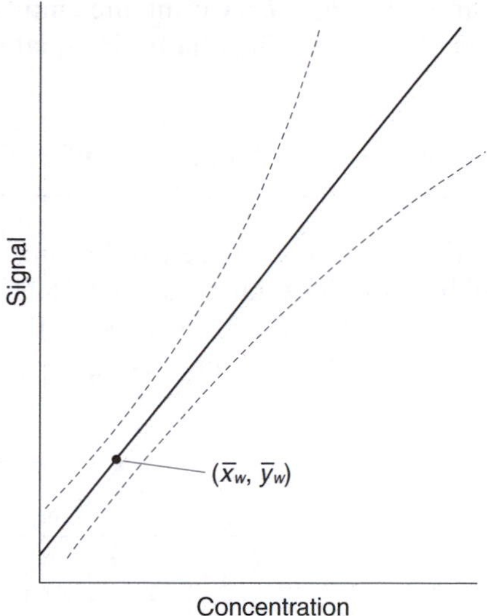

Figure 5.13 General form of the confidence limits for a concentration determined using a weighted regression line.

interval for the less concentrated of the two samples is smaller than in the unweighted regression calculation, while for the more concentrated sample the opposite is true. All these results accord much more closely with the reality of a calibration experiment than do the results of the unweighted regression calculation.

In addition, weighted regression methods may be essential when a straight line graph is obtained by algebraic transformations of an intrinsically curved plot (see Section 5.13). Computer programs for weighted regression calculations are now available, mainly through the more advanced statistical software products, and this should encourage the more widespread use of this method.

5.11 Intersection of two straight lines

A number of problems in analytical science are solved by plotting two straight line graphs from the experimental data and determining the point of their intersection. Common examples include potentiometric and conductimetric titrations, the determination of the composition of metal–chelate complexes, and studies of ligand–protein and similar bio-specific binding interactions. If the equations of the two (unweighted) straight lines, $y_1 = a_1 + b_1x_1$ and $y_2 = a_2 + b_2x_2$ (with n_1 and n_2 points respectively) are known, then the x-value of their intersection, x_I, is easily shown to be given by:

$$\text{Intersection point:} \quad x_I = \frac{\Delta a}{\Delta b} \tag{5.18}$$

where $\Delta a = a_1 - a_2$ and $\Delta b = b_2 - b_1$. Confidence limits for this x_I value are given by the two roots of the following quadratic equation:

$$x_I^2(\Delta b^2 - t^2 s_{\Delta b}^2) - 2x_I(\Delta a \Delta b - t^2 s_{\Delta a \Delta b}) + (\Delta a^2 - t^2 s_{\Delta a}^2) = 0 \quad (5.19)$$

The value of t used in this equation is chosen at the appropriate P-level and at $n_1 + n_2 - 4$ degrees of freedom. The standard deviations in equation (5.19) are calculated on the assumption that the $s_{y/x}$ values for the two lines, $s_{(y/x)1}$ and $s_{(y/x)2}$, are sufficiently similar to be pooled using an equation analogous to equation (3.3):

$$s_{(y/x)p}^2 = \frac{(n_1 - 2)s_{(y/x)1}^2 + (n_2 - 2)s_{(y/x)2}^2}{n_1 + n_2 - 4} \quad (5.20)$$

After this pooling process we can write:

$$s_{\Delta b}^2 = s_{(y/x)p}^2 \left\{ \frac{1}{\sum_i (x_{i1} - \bar{x}_1)^2} + \frac{1}{\sum_i (x_{i2} - \bar{x}_2)^2} \right\} \quad (5.21)$$

$$s_{\Delta a}^2 = s_{(y/x)p}^2 \left\{ \frac{1}{n_1} + \frac{1}{n_2} + \frac{\bar{x}_1^2}{\sum_i (x_{i1} - \bar{x}_1)^2} + \frac{\bar{x}_2^2}{\sum_i (x_{i2} - \bar{x}_2)^2} \right\} \quad (5.22)$$

$$s_{\Delta a \Delta b} = s_{(y/x)p}^2 \left\{ \frac{\bar{x}_1}{\sum_i (x_{i1} - \bar{x}_1)^2} + \frac{\bar{x}_2}{\sum_i (x_{i2} - \bar{x}_2)^2} \right\} \quad (5.23)$$

These equations seem formidable, but if a spreadsheet such as Excel is used to obtain the equations of the two lines, the point of intersection can be determined at once. The $s_{y/x}$ values can then be pooled, $s_{\Delta a}^2$, etc. calculated, and the confidence limits found using the program's equation solving capabilities.

5.12 ANOVA and regression calculations

When the least-squares criterion is used to determine the best straight line through a single set of data points there is one unique solution, so the calculations involved are relatively straightforward. However, when a curved calibration plot is calculated using the same criterion this is no longer the case: a least-squares curve might be described by polynomial functions ($y = a + bx + cx^2 + \ldots$) containing different numbers of terms, a logarithmic or exponential function, or in other ways. So we need a method which helps us to choose the best way of plotting a curve from amongst the many that are available. Analysis of variance (ANOVA) provides such a method in all cases where we maintain the assumption that the errors occur only in the y-direction. In such situations there are two sources of y-direction variation in a calibration plot. The first is the variation *due to regression*, i.e. due to the relationship between the instrument signal, y, and the analyte concentration, x. The second is the random experimental error in the y-values, which is called the variation *about regression*. As we have seen in Chapter 3, ANOVA is a powerful method for separating two sources of variation in such situations. In regression problems, the average of the y-values of the calibration points, $\bar{y}$, is important in defining these sources of variation. Individual values of y_i differ from $\bar{y}$ for the two reasons given above. ANOVA is applied to separating the two sources of variation by using the relationship that the total sum of squares (SS) about $\bar{y}$ is equal to the SS due to regression plus the SS about regression:

Additive sums of squares: $$\sum_i (y_i - \bar{y})^2 = \sum_i (\hat{y}_i - \bar{y})^2 + \sum_i (y_i - \hat{y}_i)^2 \qquad (5.24)$$

The total sum of squares, i.e. the left-hand side of equation (5.24), is clearly fixed once the experimental y_i values have been determined. A line fitting these experimental points closely will be obtained when the variation *due to regression* (the first term on the right-hand side of equation (5.24)) is as large as possible. The variation about regression (also called the *residual SS* as each component of the right-hand term in the equation is a single *residual*) should be as small as possible. The method is quite general and can be applied to straight line regression problems as well as to curvilinear regression. Table 5.1 showed the Excel output for a linear plot used to compare two analytical methods, including an ANOVA table set out in the usual way. The total number of degrees of freedom (19) is, as usual, one less than the number of measurements (20), as the residuals always add up to zero. For a straight line graph we have to determine only one coefficient (b) for a term that also contains x, so the number of degrees of freedom due to regression is 1. Thus there are $(n - 2) = 18$ degrees of freedom for the residual variation. The mean square (MS) values are determined as in previous ANOVA examples, and the F-test is applied to the two mean squares as usual. The F-value obtained is very large, as there is an obvious relationship between x and y, so the regression MS is much larger than the residual MS.

The Excel output also includes 'multiple R', which as previously noted is in this case equal to the correlation coefficient, r, the standard error ($= s_{y/x}$), and the further terms 'R square' and 'adjusted R square', usually abbreviated R'^2. The two latter statistics are given by Excel as decimals, but are often given as percentages instead. They are defined as follows:

$$R^2 = \text{SS due to regression/total SS} = 1 - (\text{residual SS/total SS}) \qquad (5.25)$$

$$R'^2 = 1 - (\text{residual MS/total MS}) \qquad (5.26)$$

In the case of a straight line graph, R^2 is equal to r^2, the square of the correlation coefficient, i.e. the square of 'multiple R'. The applications of R^2 and R'^2 to problems of curve-fitting will be discussed below.

5.13 Curvilinear regression methods – Introduction

In many instrumental analysis methods the instrument response is proportional to the analyte concentration over substantial concentration ranges. The simplified calculations that result encourage analysts to take significant experimental precautions to achieve such linearity. Examples of such precautions include the control of the emission line width of a hollow-cathode lamp in atomic absorption spectrometry, and the size and positioning of the sample cell to minimize inner filter artefacts in molecular fluorescence spectrometry. However many analytical methods (e.g. immunoassays and similar competitive binding assays) produce calibration plots that are intrinsically curved. Particularly common is the situation where the calibration plot is linear (or approximately so) at low analyte concentrations, but becomes curved at higher analyte levels. When curved calibration plots are obtained we still need answers to the questions listed in Section 5.2, but those questions will pose rather more formidable statistical problems than occur in linear calibration experiments.

The first question to be examined is – how do we detect curvature in a calibration plot? That is, how do we distinguish between a plot that is best fitted by a straight line, and one that is best fitted by a gentle curve? Since the degree of curvature may be small, and/or occur over only part of the plot, this is not a straightforward question. Moreover, despite its widespread use for testing the goodness of fit of linear graphs, the product–moment correlation coefficient (r) is of little value in testing for curvature: we have seen (Section 5.3) that lines with obvious curvature may still give very high r values. An analyst would naturally hope that any test for curvature could be applied fairly easily in routine work without extensive calculations. Several such tests are available, based on the use of the y-residuals on the calibration plot.

We have seen (Section 5.5) that a y-residual, $y_i - \hat{y}_i$, represents the difference between an experimental value of y and the $\hat{y}$ value calculated from the regression equation at the same value of x. If a linear calibration plot is appropriate, and if the random errors in the y-values are normally distributed, the residuals themselves should be normally distributed about the value of zero. If this turns out not to be true in practice, then we must suspect that the fitted regression line is not of the correct type. In the worked example given in Section 5.5 the y-residuals were shown to be +0.58, −0.38, −0.24, −0.50, +0.34, +0.18, and +0.02. These values sum to zero (allowing for possible rounding errors, this must always be true), and are approximately symmetrically distributed about 0. Although it is impossible to be certain, especially with small numbers of data points, that these residuals are normally distributed, there is certainly no contrary evidence in this case, i.e. no evidence to support a non-linear calibration plot. As previously noted Minitab and other statistics packages provide extensive information, including graphical displays, on the sizes and distribution of residuals.

A second test suggests itself on inspection of the *signs* of the residuals given above. As we move along the calibration plot, i.e. as x increases, positive and negative residuals will be expected to occur in random order if the data are well fitted by a straight line. If, in contrast, we attempt to fit a straight line to a series of points that actually lie on a smooth curve, then the signs of the residuals will no longer have a random order, but will occur in *sequences* of positive and negative values. Examining again the residuals given above, we find that the order of signs is $+----+++$. To test whether these sequences of + and – residuals indicate the need for a non-linear regression line, we need to know the probability that such an order could occur by chance. Such calculations are described in the next chapter. Unfortunately the small number of data points makes it quite likely that these and other sequences could indeed occur by chance, so any conclusions drawn must be treated with caution. The choice between straight line and curvilinear regression methods is therefore probably best made by using the curve-fitting techniques outlined in the next section.

In the situation where a calibration plot is linear over part of its range and curved elsewhere, it is of great importance to be able to establish the range over which linearity can be assumed. Approaches to this problem are outlined in the following example.

EXAMPLE 5.13.1

Investigate the linear calibration range of the following fluorescence experiment.

Fluorescence intensity	0.1	8.0	15.7	24.2	31.5	33.0
Concentration, μg ml^{-1}	0	2	4	6	8	10

Inspection of the data shows that the part of the graph near the origin corresponds rather closely to a straight line with a near-zero intercept and a slope of about 4. The fluorescence of the 10 μg ml^{-1} standard solution is clearly lower than would be expected on this basis, and there is some possibility that the departure from linearity has also affected the fluorescence of the 8 μg ml^{-1} standard. We first apply (unweighted) linear regression calculations to all the data. Application of the methods of Sections 5.3 and 5.4 gives the results $a = 1.357$, $b = 3.479$ and $r = 0.9878$. Again we recall that the high value for r may be deceptive, though it may be used in a comparative sense (see below). The y-residuals are found to be −1.257, −0.314, +0.429, +1.971, +2.314, and −3.143, with the sum of squares of the residuals equal to 20.981. The trend in the values of the residuals suggests that the last value in the table is probably outside the linear range.

We confirm this suspicion by applying the linear regression equations to the first five points only. This gives $a = 0.100$, $b = 3.950$ and $r = 0.9998$. The slope and intercept are much closer to the values expected for the part of the graph closest to the origin, and the r value is higher than in the first calculation. The residuals of the first five points from this second regression equation are 0, 0, −0.2, +0.4 and −0.2, with a sum of squares of only 0.24. Use of the second regression equation shows that the fluorescence expected from a 10 μg ml^{-1} standard is 39.6, i.e. the residual is −6.6. Use of a *t*-test (Chapter 3) would show that this last residual is significantly greater than the average of the other residuals: alternatively a test could be applied (Section 3.7) to demonstrate that it is an 'outlier' amongst the residuals (see also Section 5.15 below). In this example, such calculations are hardly necessary: the enormous residual for the last point, coupled with the very low residuals for the other five points and the greatly reduced sum of squares, confirms that the linear range of the method does not extend as far as 10 μg ml^{-1}. Having established that the last data point can be excluded from the linear range, we can repeat the process to study the point (8, 31.5). We do this by calculating the regression line for only the first four points in the table, with the results $a = 0$, $b = 4.00$, $r = 0.9998$. The correlation coefficient value suggests that this line is about as good a fit of the points as the previous one, in which five points were used. The residuals for this third calculation are +0.1, 0, −0.3, and +0.2, with a sum of squares of 0.14. With this calibration line the *y*-residual for the 8 μg ml^{-1} solution is −0.5: this value is larger than the other residuals but probably not by a significant amount. It can thus be concluded that it is reasonably safe to include the point (8, 31.5) within the linear range of the method. In making a marginal decision of this kind, the analytical chemist will take into account the accuracy required in the results, and the reduced value of a method for which the calibration range is very short. The calculations described above are summarized in Figure 5.14.

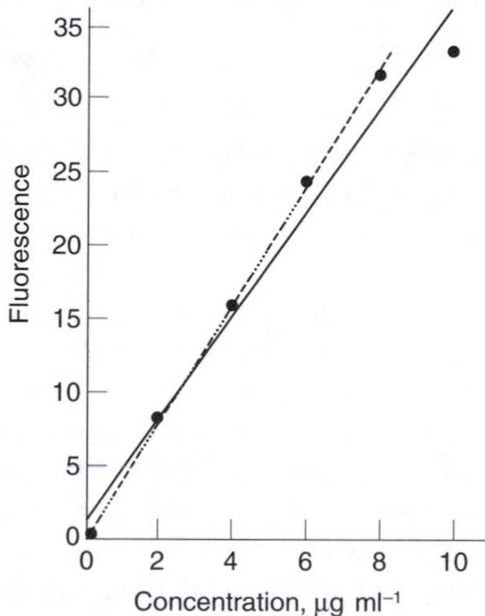

Figure 5.14 Curvilinear regression: identification of the linear range. The data in Example 5.13.1 are used; the unweighted linear regression lines through all the points (——), through the first five points only (- - - - -), and through the first four points only (· · · · ·) are shown.

Once a decision has been taken that a set of calibration points cannot be satisfactorily fitted by a straight line, the analyst can play one further card before becoming resigned to the complexities of curvilinear regression calculations. It may be possible to *transform* the data so that a non-linear relationship is changed into a linear one. Such transformations are regularly applied to the results of certain analytical methods. For example, modern software packages for the interpretation of immunoassay data frequently offer a choice of transformations: commonly used methods involve plotting log y and/or log x instead of y and x, or the use of logit functions (logit $x = \ln[x/(1-x)]$). It is important to note that the transformations may also affect the nature of the errors at different points on the calibration plot. Suppose, for example, that in a set of data of the form $y = px^q$, the magnitudes of the random errors in y are independent of x. Any transformation of the data into linear form by taking logarithms will obviously produce data in which the errors in log y are *not* independent of log x. In this case, and in any other instance where the expected form of the equation is known from theoretical considerations or from long-standing experience, it is possible to apply *weighted* regression equations (Section 5.10) to the transformed data. It may be shown that, if data of the general form $y = f(x)$ are transformed into the linear equation $Y = BX + A$, the weighting factor, w, used in equations (5.14)–(5.17) is obtained from the relationship:

$$w_i = \left\{ \frac{1}{dY_i/dy_i} \right\}^2 \tag{5.27}$$

Unfortunately, there are not many cases in analytical chemistry where the exact mathematical form of a non-linear regression equation is known with certainty (see below), so this approach is of restricted value.

It should also be noted that, in contrast to the situation described in the previous paragraph, results can be transformed to produce data that can be treated by *unweighted* methods. Data of the form $y = bx$ with y-direction errors strongly dependent on x are sometimes subjected to a log-log transformation: the errors in log y then vary less seriously with log x, so the transformed data can reasonably be studied by unweighted regression equations.

5.14 Curve fitting

In view of the difficulties that arise from transforming the data, and the increasing ease with which curves can be calculated to fit a set of calibration points, curvilinear regression methods are now relatively common in analytical chemistry. It is important to realize that the curved calibration plots encountered in practice often arise from the superposition of two or more physical or chemical phenomena. Thus in molecular fluorescence spectrometry, signal vs. concentration plots will often be approximately linear in very dilute solution, but will show increasing (negative) curvature at higher concentrations because of (a) optical artefacts (inner filter effects),

(b) molecular interactions (e.g. quenching, excimer formation), and (c) the failure of the algebraic assumptions on which a linear plot is predicted. Effects (a)–(c) are independent of one another, so many curves of different shapes may appear in practice. This example shows why calibration curves of a known and predictable form are so rarely encountered in analytical work (see above). Thus the analyst has little *a priori* guidance on which of the many types of equation that generate curved plots should be used to fit the calibration data in a particular case. In practice, much the most common strategy is to fit a curve which is a polynomial in x, i.e. $y = a + bx + cx^2 + dx^3 + \ldots$. The mathematical problems to be solved are then (i) how many terms should be included in the polynomial, and (ii) what values must be assigned to the coefficients a, b, etc? Computer software packages which address these problems are normally iterative: they fit first a straight line, then a quadratic curve, then a cubic curve, and so on, to the data, and present to the user the information needed to decide which of these equations is the most suitable. In practice quadratic or cubic equations are often entirely adequate to provide a good fit to the data: polynomials with many terms are almost certainly physically meaningless and do not significantly improve the analytical results. In any case, if the graph has n calibration points, the largest polynomial permissible is that of order $(n - 1)$.

To decide whether (for example) a quadratic or a cubic curve is the best fit to a calibration data set we can use the ANOVA methods introduced in Section 5.12. ANOVA programs generate values for R^2, the **coefficient of determination**. Equation (5.25) shows that, as the least squares fit of a curve (or straight line) to the data points improves, the value of R^2 will get closer to 1 (or 100%). It would thus seem that we have only to calculate R^2 values for the straight-line, quadratic, cubic, etc. equations, and cease our search when R^2 no longer increases. Unfortunately it turns out that the addition of another term to the polynomial always increases R^2, even if only by a small amount. ANOVA programs thus provide R'^2 ('adjusted R^2') values (equation (5.26)), which utilize mean squares (MS) rather than sums of squares. The use of R'^2 takes into account that the number of residual degrees of freedom in the polynomial regression (given by $[n - k - 1]$ where k is the number of terms in the regression equation containing a function of x) changes as the order of the polynomial changes. As the following example shows, R'^2 is always smaller than R^2.

EXAMPLE 5.14.1

In an instrumental analysis the following data were obtained (arbitrary units).

Concentration	0	1	2	3	4	5	6	7	8	9	10
Signal	0.2	3.6	7.5	11.5	15.0	17.0	20.4	22.7	25.9	27.6	30.2

Fit a suitable polynomial to these results, and use it to estimate the concentrations corresponding to signal of 5, 16 and 27 units.

Even a casual examination of the data suggests that the calibration plot should be a curve, but it is instructive nonetheless to calculate the least-squares straight line through the points using the method described in Section 5.4. This line turns out to have the equation $y = 2.991x + 1.555$. The ANOVA table for the data has the following form:

Source of variation	*Sum of squares*	*d.f.*	*Mean square*
Regression	984.009	1	984.009
Residual	9.500	9	1.056
Total	993.509	10	99.351

As already noted the number of degrees of freedom (d.f.) for the variation due to regression is equal to the number of terms (k) in the regression equation containing x, x^2, etc. For a straight line, k is 1. There is only one constraint in the calculation (viz. that the sum of the residuals is zero, see above), so the total number of degrees of freedom is $n - 1$. Thus the number of degrees of freedom assigned to the residuals is $(n - k - 1) = (n - 2)$ in this case. From the ANOVA table R^2 is given by 984.009/993.509 = 0.99044, i.e. 99.044%. An equation which explains over 99% of the relationship between x and y seems quite satisfactory but, just as is the case with the correlation coefficient, r, we must use great caution in interpreting absolute values of R^2: it will soon become apparent that a quadratic curve provides a much better fit for the data. We can also calculate the R'^2 value from equation (5.26): it is given by $(1 - [1.056/99.351]) = 0.98937$, i.e. 98.937%.

As always an examination of the residuals usually provides valuable information on the success of a calibration equation. In this case the residuals are as follows:

x	y_i	$\hat{y}_i$	*y-residual*
0	0.2	1.0	−1.4
1	3.6	4.5	−0.9
2	7.5	7.5	0
3	11.5	10.5	1.0
4	15.0	13.5	1.5
5	17.0	16.5	0.5
6	20.4	19.5	0.9
7	22.7	22.5	0.2
8	25.9	25.5	0.4
9	27.6	28.5	−0.9
10	30.2	31.5	−1.3

In this table, the numbers in the two right-hand columns have been rounded to one decimal place for simplicity. The trend in the signs and magnitudes of the residuals, which are negative at low x-values, rise to a positive maximum, and then return to negative values, is a sure sign that a straight line is not a suitable fit for the data.

When the data are fitted by a curve of quadratic form the equation turns out to be $y = 0.086 + 3.970x - 0.098x^2$, and the ANOVA table takes the form:

Source of variation	*Sum of squares*	*d.f.*	*Mean square*
Regression	992.233	2	494.116
Residual	1.276	8	0.160
Total	993.509	10	99.351

Note that the number of degrees of freedom for the regression and residual sources of variation have now changed in accordance with the rules described above, but that the total variation is naturally the same as in the first ANOVA table. Here R^2 is $992.233/993.509 = 0.99872$, i.e. 99.872%. This figure is noticeably higher than the value of 99.044% obtained from the linear plot, and the R'^2 value is also higher at $(1 - [0.160/99.3511]) = 0.99839$, i.e. 99.839%. When the *y*-residuals are calculated, their signs (in increasing order of *x*-values) are $+ - - + + - + - + - +$. There is no obvious trend here, so on all grounds we must prefer the quadratic over the linear fit.

Lastly we repeat the calculation for a cubit fit. Here, the best-fit equation is $y = -0.040 + 4.170x - 0.150x^2 + 0.0035x^3$. The cubic coefficient is very small indeed, so it is questionable whether this equation is a significantly better fit than the quadratic one. The R^2 value is, inevitably, slightly higher than that for the quadratic curve (99.879% compared with 99.872%), but the value of R'^2 is slightly *lower* than the quadratic value at 99.827%. The order of the signs of the residuals is the same as in the quadratic fit. As there is no value in including unnecessary terms, we can be confident that a quadratic fit is satisfactory in this case.

When the above equations are used to estimate the concentrations corresponding to instrument signals of 5, 16 and 27 units, the results (*x*-values in arbitrary units) are:

	Linear	*Quadratic*	*Cubic*
$y = 5$	1.15	1.28	1.27
$y = 16$	4.83	4.51	4.50
$y = 27$	8.51	8.61	8.62

As expected, the differences between the concentrations calculated from the quadratic and cubic equations are insignificant, so the quadratic equation is used for simplicity.

We noted earlier in this section that non-linear calibration graphs often result from the simultaneous occurrence of a number of physicochemical and/or mathematical phenomena, so it is sensible to assume that no single mathematical function may be able to describe the calibration curve entirely satisfactorily. It thus seems logical to try to fit the points to a curve that consists of several linked sections whose mathematical form may be different. This is the approach now used with increasing frequency through the application of **spline functions**. **Cubic splines** are most commonly used in practice, i.e. the final curve is made up of a series of linked sections of cubic form. These sections must clearly form a continuous curve at their junctions ('knots'), so the first two derivatives of each curve at any knot must be identical. Several methods have been used for estimating both the number of knots and the equations of the curves joining them: these techniques are too advanced to be considered in detail here, but many commercially available statistics software packages now provide such facilities. The spline function approach has been applied successfully to a variety of analytical methods, including gas–liquid chromatography, competitive binding immunoassays and similar receptor-based methods, and atomic-absorption spectrometry.

It is legitimate to ask whether, in the case of a calibration plot whose curvature is not too severe, we could take the spline idea to its simplest conclusion, and plot the curve as a series of straight lines joining successive points. This method is of course entirely non-rigorous, and would not provide any information on the precision with which x-values can be determined. However, its value as a simple initial data analysis (IDA) method (see Chapter 6) is indicated by applying it to the data in the above example. For y-values of 5, 16 and 27 this method of linear interpolation between successive points gives x-values of 1.36, 4.50 and 8.65 units respectively. Comparison with the above table shows that these results, especially the last two, would be quite acceptable for many purposes.

5.15 Outliers in regression

In this section we return to a problem already discussed in Chapter 3, the occurrence of outliers in our data. These anomalous results inevitably arise in calibration experiments, just as they occur in replicate measurements, but it is rather harder to deal with them in regression statistics. One difficulty is that, although the individual y_i-values in a calibration experiment are assumed to be independent of one another, the residuals $(y_i - \hat{y}_i)$ are not independent of one another, as their sum is always zero. It is therefore not normally permissible to take the residuals as if they were a conventional set of replicate measurements, and apply (for example) a Q-test to identify any outliers. (If the number of y_i-values is large, a condition not generally met in analytical work, this prohibition can be relaxed.)

How then do we identify outliers in a typical calibration experiment? First we note that, in cases where an obvious error such as a transcription mistake or an instrument malfunction has occurred it is natural and permissible to reject the resulting measurement (and, if possible, to repeat it). If there are suspect measurements for which there are no obvious sources of error, we must return to a study of the residuals. Most computer programs handling regression data provide residual diagnostics routines (see above). Some of these are simple, including plots of the individual residuals against y_i-values

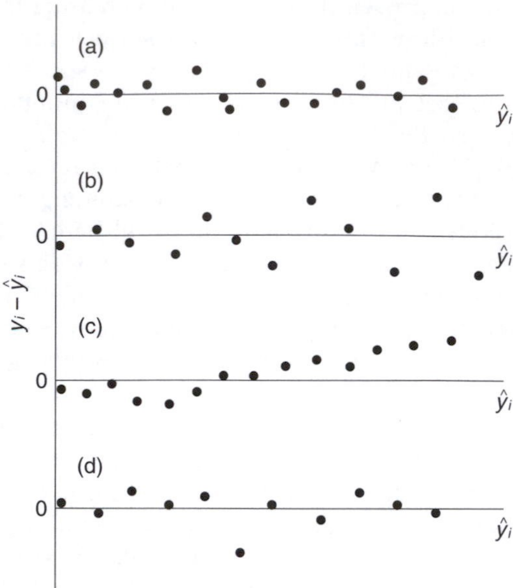

Figure 5.15 Residual plots in regression diagnostics: (a) satisfactory distribution of residuals; (b) the residuals tend to grow as y_i grows, suggesting that a weighted regression plot would be suitable; (c) the residuals show a trend, first becoming more negative, then passing through zero, and then becoming more positive as y_i increases, suggesting that a (different) curve should be plotted; and (d) a satisfactory plot, except that y_6 might be an outlier.

(Figure 5.15). Such plots would normally be expected to show that, if the correct calibration model been used, the residuals remain roughly uniform in size as y_i increases, and normally distributed about zero. The figure also illustrates cases where the y-direction errors increase with y_i (Section 5.10), and where the wrong regression equation has been used (Sections 5.11 and 5.12). Similarly, the y-residuals can be plotted against time if instrument drift or any other time-dependent effect is suspected. These plots show up suspect values very clearly, but do not provide criteria that can be immediately used to reject or accept them. Moreover, they are of limited value in many analytical chemistry experiments, where the number of calibration points is often small.

Some simple numerical criteria have been used in computer software to identify possible outliers. Some packages 'flag' calibration points where the y-residual is more than twice (or some other multiple of) the value of $s_{y/x}$. Several more advanced methods have been developed, of which the best known is the estimation for each point of **Cook's distance**, first proposed in 1977. This statistic is provided routinely by several advanced suites of statistical software, though a full appreciation of its significance requires a knowledge of matrix algebra. Cook's distance is an example of an *influence function*, i.e. it measures the effect that rejecting the calibration point in question would have on the regression coefficients a and b.

Finally we note that, just as in the treatment of outliers in replicate measurements, non-parametric and robust methods can be very effective in handling outliers in regression: robust regression methods have proved particularly popular in recent years. These topics are covered in the next chapter.

Bibliography

Draper, N. R. and Smith, H. 1998. *Applied Regression Analysis.* 3rd Edn. Wiley, New York. (An established work with comprehensive coverage of many aspects of regression and correlation problems.)

Edwards, A. L. 1984. *An Introduction to Linear Regression and Correlation.* 2nd Edn. W. H. Freeman, New York. (Clearly written treatment, with a good introduction to matrix algebra.)

Kleinbaum, D. G., Kupper, L. L. and Muller, K. E. 1988. *Applied Regression Analysis and Other Multivariable Methods.* 2nd Edn. PWS-Kent, Boston. (Another classic in its field with a good treatment of regression diagnostics.)

Mark, H. 1991. *Principles and Practice of Spectroscopic Calibration.* Wiley, New York. (A clear explanation of the principles. The strong emphasis on near-IR spectroscopic data is not a significant disadvantage.)

Noggle, J. H. 1993. *Practical Curve-Fitting and Data Analysis.* Ellis Horwood – PTR Prentice Hall, New Jersey. (Software and data files are provided with this book, and all the examples are chemical ones.)

Orvis, W. J. 1996. *Excel for Scientists and Engineers.* 2nd Edn. Sybex, Alameda, California. (One of a number of books giving clear guidance on the use of Excel for statistical calculations.)

Snedecor, G. M. and Cochran, W. G. 1989. *Statistical Methods.* 8th Edn. Iowa State University, USA. (Gives an excellent general account of regression and correlation procedures.)

Exercises

1. In a laboratory containing polarographic equipment six samples of dust were taken at various distances from the polarograph and the mercury content of each sample was determined. The following results were obtained:

Distance from polarograph, m	1.4	3.8	7.5	10.2	11.7	15.0
Mercury concentration, ng g^{-1}	2.4	2.5	1.3	1.3	0.7	1.2

Examine the possibility that the mercury contamination arose from the polarograph.

2. The response of a colorimetric test for glucose was checked with the aid of standard glucose solutions. Determine the correlation coefficient from the following data and comment on the result.

Glucose concentration, mM	0	2	4	6	8	10
Absorbance	0.002	0.150	0.294	0.434	0.570	0.704

3. The following results were obtained when each of a series of standard silver solutions was analysed by flame atomic-absorption spectrometry.

Concentration, ng ml^{-1}	0	5	10	15	20	25	30
Absorbance	0.003	0.127	0.251	0.390	0.498	0.625	0.763

Determine the slope and intercept of the calibration plot, and their confidence limits.

4. Using the data of exercise 3, estimate the confidence limits for the silver concentrations in (a) a sample giving an absorbance of 0.456 in a single determination, and (b) a sample giving absorbance values of 0.308, 0.314, 0.347, and 0.312 in four separate analyses.

5. Estimate the limit of detection of the silver analysis from the data in exercise 3.

6. The gold content of a concentrated sea-water sample was determined by using atomic-absorption spectrometry with the method of standard additions. The results obtained were as follows:

Gold added, ng per ml of concentrated sample	0	10	20	30	40	50	60	70
Absorbance	0.257	0.314	0.364	0.413	0.468	0.528	0.574	0.635

Estimate the concentration of the gold in the concentrated sea-water, and determine confidence limits for this concentration.

7. The fluorescence of each of a series of acidic solutions of quinine was determined five times. The results are given below.

Concentration, ng ml^{-1}	0	10	20	30	40	50
Fluorescence intensity	4	22	44	60	75	104
(arbitrary units)	3	20	46	63	81	109
	4	21	45	60	79	107
	5	22	44	63	78	101
	4	21	44	63	77	105

Determine the slopes and intercepts of the unweighted and weighted regression lines. Calculate, using both regression lines, the confidence limits for the concentrations of solutions with fluorescence intensities of 15 and 90 units.

8. An ion-selective electrode (ISE) determination of sulphide from sulphate reducing bacteria was compared with a gravimetric determination. The results obtained were expressed in milligrams of sulphide.

Sample:	1	2	3	4	5	6	7	8	9	10
Sulphide (ISE method):	108	12	152	3	106	11	128	12	160	128
Sulphide (gravimetry):	105	16	113	0	108	11	141	11	182	118

Comment on the suitability of the ISE method for this sulphide determination.
(Al-Hitti, I. K., Moody, G. J. and Thomas, J. D. R. 1983. *Analyst* 108: 43)

9. In the determination of lead in aqueous solution by electrochemical atomic-absorption spectrometry with graphite-probe atomization, the following results were obtained:

Lead concentration, ng ml^{-1}	10	25	50	100	200	300
Absorbance	0.05	0.17	0.32	0.60	1.07	1.40

Investigate the linear calibration range of this experiment.
(Based on Giri, S. K., Shields, C. K., Littlejohn, D. and Ottaway, J. M. 1983. *Analyst* 108: 244)

10. In a study of the complex formed between europium (III) ions and pyridine-2,6-dicarboxylic acid (DPA), the absorbance values of solutions containing different DPA: Eu concentrations were determined, with the following results:

Absorbance	0.008	0.014	0.024	0.034	0.042	0.050	0.055	0.065	0.068	0.076
DPA : Eu	0.2	0.4	0.6	0.8	1.0	1.2	1.4	1.6	1.8	2.0

Absorbance	0.077	0.073	0.066	0.063	0.058
DPA:Eu	2.4	2.8	3.2	3.6	4.0

Use these data to determine the slopes and intercepts of two separate straight lines. Estimate their intersection point and its standard deviation, thus determining the composition of the DPA–europium complex formed.
(Based on Arnaud, N., Vaquer, E. and Georges, J. 1998. *Analyst* 123: 261)

11. In an experiment to determine hydrolysable tannins in plants by absorption spectroscopy the following results were obtained:

Absorbance	0.084	0.183	0.326	0.464	0.643
Concentration, mg ml^{-1}	0.123	0.288	0.562	0.921	1.420

Use a suitable statistics or spreadsheet program to calculate a quadratic relationship between absorbance and concentration. Using R^2 and R'^2 values, comment on whether the data would be better described by a cubic equation.
(Based on Willis, R. B. and Allen, P. R. 1998. *Analyst* 123: 435)

12. The following results were obtained in an experiment to determine spermine by high performance thin layer chromatography of one of its fluorescent derivatives.

Fluorescence intensity	36	69	184	235	269	301	327
Spermine, ng	6	18	30	45	60	75	90

Determine the best polynomial calibration curve through these points. (Based on Linares, R. M., Ayala, J. H., Afonso, A. M. and Gonzalez, V. 1998. *Analyst* 123: 725)

CHAPTER SIX

Non-parametric and robust methods

6.1 Introduction

The statistical tests developed in the previous chapters have all assumed that the data being examined follow the normal (Gaussian) distribution. Some support for this assumption is provided by the central limit theorem, which shows that the sampling distribution of the mean may be approximately normal even if the parent population has quite a different distribution. However, the theorem is not really valid for the very small data sets (often only three or four readings) frequently used in analytical work.

There are several further reasons for an interest in methods that do not require the assumption of normally distributed data. Some sets of data that are of interest to analytical chemists certainly have different distributions. For example (see Chapter 2) the concentrations of antibody in the blood sera of a group of different people can be expressed approximately as a log-normal distribution: such results are often obtained when a particular measurement is made on each member of a group of human or animal subjects. More interestingly, there is growing evidence that, even when repeated measurements are made on a *single* test material, the distribution of the results may sometimes be symmetrical but not normal: the data include more results than expected which are distant from the mean. Such **heavy-tailed** distributions may be regarded as normal distributions with the addition of outliers (see Chapter 3) arising from gross errors. Alternatively heavy-tailed data may arise from the superposition of two normal distributions with the same mean value, but with one distribution having a significantly larger standard deviation than the other. This could arise if, for example, the measurements were made by more than one individual or by using more than one piece of equipment.

This chapter introduces two groups of statistical tests for handling data that may not be normally distributed. Methods which make no assumptions about the shape of the distribution from which the data are taken are called

non-parametric or **distribution-free** methods. Many of them have the further advantage of greatly simplified calculations: with small data sets some of the tests can be performed mentally. The second group of methods, which has grown rapidly in use in recent years, is based on the belief that the underlying population distribution may indeed be normal (or have some other well-defined form), but with the addition of data such as outliers that may distort this distribution. These **robust** techniques will naturally be appropriate in the cases of heavy-tailed distributions described above. They differ from non-parametric methods in another respect: they often involve iterative calculations that would be lengthy or complex without a computer, and their rise in popularity certainly owes much to the universal availability of personal computers.

6.2 The median: initial data analysis

In previous chapters we have used the arithmetic mean or average as the 'measure of central tendency' or 'measure of location' of a set of results. This is logical enough when the (symmetrical) normal distribution is assumed, but in non-parametric statistics, the **median** is usually used instead. In order to calculate the median of *n* observations, we arrange them in ascending order: in the unlikely event that *n* is very large, this sorting process can be performed very quickly by programs available for most computers.

> The median is the value of the $\frac{1}{2}(n+1)$th observation if n is odd: and the average of the $\frac{1}{2}n$th and the $(\frac{1}{2}n+1)$th observations if n is even.

Determining the median of a set of experimental results usually requires little or no calculation. Moreover, in many cases it may be a more realistic measure of central tendency than the arithmetic mean.

EXAMPLE 6.2.1

Determine the mean and the median for the following four titration values.

25.01, 25.04, 25.06, 25.21 ml

It is easy to calculate that the mean of these four observations is 25.08 ml, and that the median – in this case the average of the second and third values, the observations already being in numerical order – is 25.05 ml. The mean is greater than any of the three closely-grouped values (25.01. 25.04 and 25.06 ml) and may thus be a less realistic measure of central tendency than the median. Instead of calculating the median we could use the methods of Chapter 3 to test the value 25.21 as a possible outlier, and determine the mean according to the result obtained, but this approach involves extra calculation and assumes that the data come from a normal population.

This simple example illustrates one valuable property of the median: it is unaffected by outlying values. Confidence limits (see Chapter 2) for the median can be estimated with the aid of the binomial distribution. This calculation can be performed even when the number of measurements is small, but is not likely to be required in analytical chemistry, where the median is generally used only as a rapid measure of an average. The reader is referred to the bibliography for further information.

In non-parametric statistics the usual measure of dispersion (replacing the standard deviation) is the **interquartile range.** As we have seen, the median divides the sample of measurements into two equal halves: if each of these halves is further divided into two the points of division are called the **upper** and **lower quartiles.** Several different conventions are used in making this calculation, and the interested reader should again consult the bibliography. The interquartile range is not widely used in analytical work, but various statistical tests can be performed on it.

The median and the interquartile range of a set of measurements are just two of the statistics which feature strongly in **initial data analysis** (IDA), often also called **exploratory data analysis** (EDA). This is an aspect of statistics that has grown rapidly in popularity in recent years. One reason for this is, yet again, the ability of modern computers and dedicated software to present data almost instantly in a wide range of graphical formats: as we shall see, such pictorial representations form an important element of IDA. A second element in the rising importance of IDA is the increasing acceptance of statistics as a practical and pragmatic subject not necessarily restricted to the use of techniques whose theoretical soundness is unquestioned: some IDA methods seem almost crude in their principles, but have nonetheless proved most valuable.

> The main advantage of IDA methods is their ability to indicate which (if any) further statistical methods are most appropriate to a given data set.

Several simple presentation techniques are of immediate help. We have already noted (see Chapters 1 and 3) the use of **dot-plots** in the illustration of small data sets. These plots help in the visual identification of outliers and other unusual features of the data. The following example further illustrates their value.

EXAMPLE 6.2.2

In an experiment to determine whether Pb^{2+} ions interfered with the enzymatic determination of glucose in various foodstuffs, nine food materials were treated with a 0.1 mM solution of Pb(II), while four other materials (the control group) were left untreated. The rates (arbitrary units) of the enzyme catalysed reaction were then measured for each food and corrected for the different amounts of glucose known to be present. The results were:

Treated foods	21	1	4	26	2	27	11	24	21
Controls	22	22	32	23					

Comment on these data.

Written out in two rows as above, the data do not convey much immediate meaning, and an unthinking analyst might proceed straight away to perform a *t*-test (Chapter 3), or perhaps one of the non-parametric tests described below, to see if the two sets of results are significantly different. But when the data are presented as two dot plots, or as a single plot with the two sets of results given separate symbols, it is apparent that the results, while interesting, are so inconclusive that little can be deduced from them without further measurements (Figure 6.1).

The medians of the two sets of data are similar: 21 for the treated foods and 22.5 for the controls. But the range of reaction rates for the Pb(II) treated materials is enormous, with the results apparently falling into at least two groups: five of the foods seem not to be affected by the lead (perhaps because in these cases Pb(II) is complexed by components other than the enzyme in question), while three others show a large inhibition effect (i.e. the reaction rate is much reduced), and another lies somewhere in between these two extremes. There is the further problem that one of the control group results is distinctly different from the rest, and might be considered an outlier (Chapter 3). In these circumstances it seems most unlikely that a conventional significance test will reveal chemically useful information: the use of the simplest IDA method has guided us away from thoughtless significance testing and (as so often happens) towards more experimental measurements.

Another simple data representation technique, of greater value when rather larger samples are studied, is the **box-and-whisker plot**. In its normal form such a diagram consists of a rectangle (the box) with two lines (the whiskers) extending from opposite edges of the box, and a further line in the box, crossing it parallel to the same edges. The ends of the whiskers indicate the range of the data, the edges of the box from which the whiskers protrude represent the upper and lower quartiles, and the line crossing the box represents the median of the data (Figure 6.2).

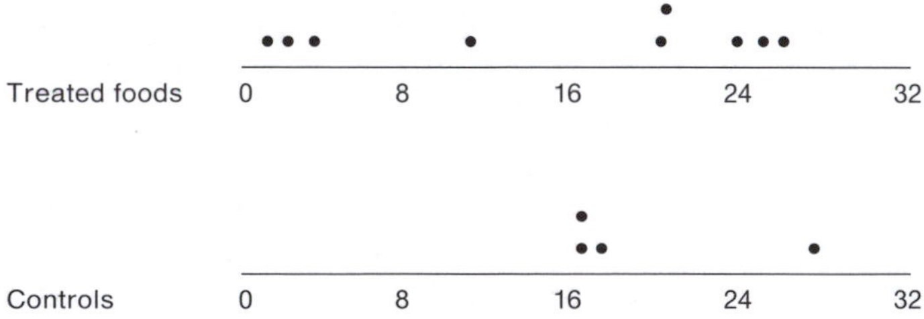

Figure 6.1 Dot-plots for Example 6.2.2.

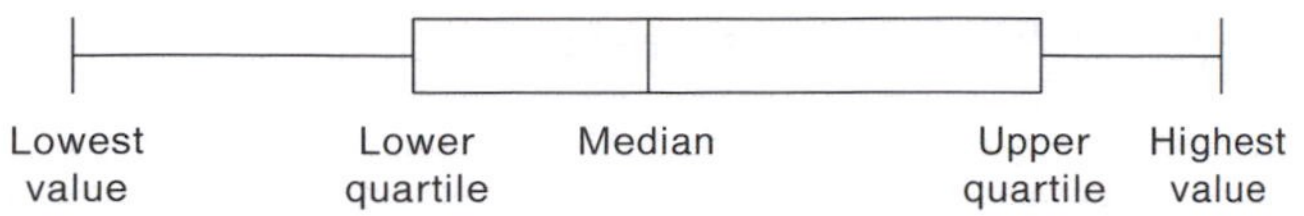

Figure 6.2 Box-and-whisker plot.

The box-and-whisker plot, accompanied by a numerical scale, is a graphical representation of the **five-number summary**, i.e. the data set is described by its extremes, its lower and upper quartiles, and its median. The plot shows at a glance the spread and the symmetry of the data.

Some computer programs enhance the data presentation by identifying possible outliers separately. In such cases outliers are often defined as data points which are lower than the lower quartile, or higher than the upper quartile, by more than 1.5 times the interquartile range. The whiskers then only extend to these upper and lower limits or **fences** and outlying data are shown as separate points. (These refinements are not shown in Figure 6.2).

EXAMPLE 6.2.3

The levels of a blood plasma protein in 20 men and 20 women (mg 100 ml^{-1}) were found to be:

Men	3	2	1	4	3	2	9	13	11	3
	18	2	4	6	2	1	8	5	1	14
Women	6	5	2	1	7	2	2	11	2	1
	1	3	11	3	2	3	2	1	4	8

What information can be gained about any differences between the levels of this protein in men and women?

As in the previous example, the data as presented convey very little, but the use of two box-and-whisker plots or five-number summaries is very revealing. The five-number summaries are:

	Min.	*Lower quartile*	*Median*	*Upper quartile*	*Max.*
Men	1	2	3.5	8.5	18
Women	1	2	2.5	5.5	11

It is left as a simple sketching exercise for the reader to show that (a) the distributions are very skewed in both men and women, so statistical methods that assume a normal distribution are not appropriate (as we have seen this is often true when a single measurement is made on a number of *different* subjects, particularly when the latter are living organisms); (b) the median concentrations for men and women are similar: and (c) the range of values is considerably greater for men than for women. The conclusions suggest that we might apply the Siegel–Tukey test (see Section 6.6) to see whether the greater variation in protein levels amongst men is significant.

Table 6.1 Levels of pp-DDT in 30 butter bean specimens (mg kg^{-1})

0.03	0.05	0.08	0.08	0.10	0.11	0.18	0.19	0.20	0.20
0.22	0.22	0.23	0.29	0.30	0.32	0.34	0.40	0.47	0.48
0.55	0.56	0.58	0.64	0.66	0.78	0.78	0.86	0.89	0.96

While it is usual for analysts to handle relatively small sets of data there are occasions when a larger set of measurements is to be examined. Examples occur in the areas of clinical and environmental analysis, where in many instances there are large natural variations in analyte levels. Table 6.1 shows, in numerical order, the levels of a pesticide in 30 samples of butter beans. The individual values range from 0.03 to 0.96 mg kg^{-1}. They might be expressed as a histogram. This would show that, for example, there are four values in the range 0–0.095 mg kg^{-1}, four in the range 0.095–0.195 mg kg^{-1}, and so on. But a better IDA method uses a **stem and leaf diagram**, as shown in Figure 6.3.

The left-hand column of figures – the stem – shows the first significant digit for each measurement, while the remaining figures in each row – the leaves – provide the second significant digit. The length of the rows thus correspond to the lengths of the bars on the corresponding histogram, but the advantage of the stem and leaf diagram is that it retains the value of each measurement. The leaves use only whole numbers, so some indication of the scale used must always be given. In this case a key is used to provide this information. The Minitab software package provides facilities for stem and leaf diagrams.

> In summary, IDA methods are simple, readily handled by personal computers, and most valuable in indicating features of the data not apparent on initial inspection. They are helpful in deciding the most suitable significance tests or other statistical procedures to be adopted in further work, and sometimes even in suggesting that statistics has no further role to play until more data are obtained.

```
0 | 3 5 8 8
1 | 0 1 8 9
2 | 0 0 2 2 3 9
3 | 0 2 4
4 | 0 7 8
5 | 5 6 8
6 | 4 6
7 | 8 8
8 | 6 9
9 | 6            Key: 1|1 = 0.11 mg kg^-1
```

Figure 6.3 Stem and leaf diagram for data from Table 6.1.

They can of course be extended to the area of calibration and other regression techniques: the very crude method of plotting a curved calibration graph suggested at the end of the previous chapter can be regarded as an IDA approach. Numerous techniques are described in the books by Chatfield and by Velleman and Hoaglin listed in the bibliography at the end of this chapter.

6.3 The sign test

The sign test is amongst the simplest of all non-parametric statistical methods, and was first discussed in the early eighteenth century. It can be used in a number of ways, the simplest of which is demonstrated by the following example.

EXAMPLE 6.3.1

A pharmaceutical preparation is claimed to contain 8% of a particular component. Successive batches were found in practice to contain 7.3, 7.1, 7.9, 9.1, 8.0, 7.1, 6.8 and 7.3% of the constituent. Are these results consistent with the manufacturer's claim?

In Chapter 3 (Section 3.2) it was shown that such problems could be tackled by using the t-test after calculation of the mean and standard deviation of the experimental data. The t-test assumes, however, that the data are normally distributed. The sign test avoids such an assumption, and is much easier to perform. The same underlying principles are used as in other significance tests: a null hypothesis is established, the probability of obtaining the experimental results is determined, and the null hypothesis is rejected if this probability is less than a certain critical level. Here the null hypothesis is that the data come from a population with a median value of 8.0% of the constituent. This postulated median is subtracted from each experimental value in turn, and the **sign** of each result is considered. Values equal to the postulated median are ignored *entirely*. In this case, therefore, we effectively have seven experimental values, six of them lower than the median and hence giving minus signs, and one higher than the median and hence giving a plus sign. To test whether this preponderance of minus signs is significant we use the binomial theorem. This theorem shows that the probability of r out of n signs being minus is given by

$$P(r) = {}^{n}C_{r}p^{r}q^{(n-r)} \tag{6.1}$$

where ${}^{n}C_{r}$ is the number of combinations of r items from a total of n items, p is the probability of getting a minus sign in a single result, and q is the probability of not getting a minus sign in a single result. i.e. $q = 1 - p$. Since the median is defined so that half the experimental results lie above it and half below it, it is clear that if the median is 8.0 in this case then both p and q should be $\frac{1}{2}$. Using equation (6.1) we find that $P(6) = {}^{7}C_{6} \times (\frac{1}{2})^{6} \times \frac{1}{2} = 7/128$. Similarly we can calculate that the chance of getting seven minus signs, $P(7)$, is 1/128. Overall, therefore, the probability of getting *6 or more* negative signs in our experiment is 8/128. We are only asking, however, whether the data *differ* significantly from the postulated median. We must therefore perform a two-sided test (see Chapter 3), i.e. we must calculate the probability of obtaining six or more identical signs (i.e. ≥ 6 plus or ≥ 6 minus signs) when seven results are taken at random. This is clearly $16/128 = 0.125$. Since this value is >0.05, the critical probability level usually used, the null hypothesis, i.e. that the data come from a population with median 8.0, cannot be rejected. As in Chapter 3, it is important to note that we have not proved that the data do come from such a population; we have only concluded that such a hypothesis cannot be rejected.

It is apparent from this example that the sign test will involve the frequent use of the binomial distribution with $p = q = \frac{1}{2}$. So common is this approach to non-parametric statistics that most sets of statistical tables include the necessary data, allowing such calculations to be made instantly (see Table A.9). Moreover, in many practical situations, an analyst will always take the same number of readings or samples, and will be able to memorize easily the probabilities corresponding to the various numbers of + or − signs.

The sign test can also be used as a non-parametric alternative to the paired *t*-test (Section 3.4) to compare two sets of results for the same samples. Thus if ten samples are examined by each of two methods, A and B, we can test whether the two methods give significantly different readings by calculating for each sample [(result obtained by method A) − (result obtained by method B)]. The null hypothesis will be that the two methods do not give significantly different results – in practice this will again mean that the probability of obtaining a plus sign (or a minus sign) for each difference is 0.5. The number of plus or minus signs actually obtained can be compared with the probability derived from equation (6.1). An example of this application of the sign test is given in the exercises at the end of the chapter.

A further use of the sign test is to indicate a trend. This application is illustrated by the following example.

EXAMPLE 6.3.2

The level of a hormone in a patient's blood plasma is measured at the same time each day for 10 days. The resulting data are:

Day	1	2	3	4	5	6	7	8	9	10
Level, ng ml^{-1}	5.8	7.3	4.9	6.1	5.5	5.5	6.0	4.9	6.0	5.0

Is there any evidence for a trend in the hormone concentration?

Using parametric methods, it would be possible to make a linear regression plot of such data and test whether its slope differed significantly from zero (Chapter 5). Such an approach would assume that the errors were normally distributed, and that any trend that did occur was linear. The non-parametric approach is again simpler. The data are divided into two equal sets, the sequence being retained:

5.8	7.3	4.9	6.1	5.5
5.5	6.0	4.9	6.0	5.0

(If there is an odd number of measurements, the middle one in the time sequence is ignored.) The result for the sixth day is then subtracted from that for the first day, that for the seventh day from that for the second day, etc. The signs of the differences between the pairs of values in the five columns are determined in this way to be +, +, 0, +, +. As usual the zero is ignored completely, leaving four results, all positive. The probability of obtaining four identical signs in four trials is $2 \times (1/16) = 0.125$. (Note that a two-sided test is again used, as the trend in the hormone level might be upwards or downwards.) The null hypothesis, that there is no trend in the results, can therefore not be rejected at the $P = 0.05$ probability level.

The price paid for the extreme simplicity of the sign test is some loss of statistical power. The test does not utilize all the information offered by the data, so it is not surprising to find that it also provides less discriminating information. In later sections, non-parametric methods that do use the magnitudes of the individual results as well as their signs will be discussed.

6.4 The Wald–Wolfowitz runs test

In some instances we are interested not merely in whether observations generate positive or negative signs, but also in whether these signs occur in a random sequence. In Section 5.11, for example, we showed that if a straight line is a good fit to a set of calibration points, positive and negative residuals will occur more or less at random. By contrast, attempting to fit a straight line to a set of points that actually lie on a curve will yield non-random sequences of positive or negative signs: there might, for example, be a sequence of + signs, followed by a sequence of – signs, and then another sequence of + signs. Such sequences are technically known as **runs** – the word being used here in much the same way as when someone refers to 'a run of bad luck', or 'a run of high scores'. In the curve-fitting case, it is clear that a non-random sequence of + and – signs will lead to a smaller number of runs than a random sequence.

The Wald–Wolfowitz method tests whether the number of runs is small enough for the null hypothesis of a random distribution of signs to be rejected.

The number of runs in the experimental data is compared with the numbers in Table A.10, which refers to the $P = 0.05$ probability level. The table is entered by using the appropriate values for N, the number of + signs, and M, the number of − signs. If the experimental number of runs is *smaller* than the tabulated value, then the null hypothesis can be rejected.

EXAMPLE 6.4.1

Linear regression equations are used to fit a straight line to a set of 12 calibration points. The signs of the resulting residuals in order of increasing x value are: + + + + − − − − − − ++. Comment on whether it would be better to attempt to fit a curve to the points.

Here $M = N = 6$, and the number of runs is three. Table A.10 shows that, at the $P = 0.05$ level, the number of runs must be <4 if the null hypothesis is to be rejected. So in this instance we can reject the null hypothesis, and conclude that the sequence of + and − signs is not a random one. The attempt to fit a straight line to the experimental points is therefore unsatisfactory, and a curvilinear regression plot is indicated instead.

The Wald–Wolfowitz test can be used with any results that can be divided or converted into just two categories. Suppose, for example, that it is found that 12 successively used spectrometer light sources last for 450, 420, 500, 405, 390, 370, 380, 395, 370, 370, 420 and 430 hours. The median lifetime, in this case the average of the sixth and seventh numbers when the data are arranged in ascending order, is 400 hours. If all those lamps with lifetimes less than the median are given a minus sign, and those with longer lifetimes are given a plus sign, then the sequence becomes: + + + + − − − − − − ++. This is the same sequence as in the regression example above, where it was shown to be significantly non-random. In this case, the significant variations in lifetime might be explained if the lamps came from different batches or different manufacturers.

We may be concerned with unusually large numbers of short runs, as well as unusually small numbers of long runs. If six plus and six minus signs occurred in the order + − + − + − + − + − +− we would strongly suspect a non-random sequence. Table A.10 shows that, with $N = M = 6$, a total of 11 or 12 runs indicates that the null hypothesis of random order should be rejected, and some periodicity in the data suspected.

6.5 The Wilcoxon signed rank test

Section 6.3 described the use of the sign test. Its value lies in the minimal assumptions it makes about the experimental data. The population from which the sample is taken is not assumed to be normal, or even to be symmetrical. On the other hand a disadvantage of the sign test is that it uses so little of the information provided. The only material point is whether an individual measurement is greater than or less than the median – the size of this deviation is not used at all.

In many instances we will have every reason to believe that our measurements will be *symmetrically* distributed but will not wish to make the assumption that they are normally distributed. This assumption of symmetrical data, and the consequence that the mean and the median of the population will be equal, allows more powerful significance tests to be developed. Important advances were made by Wilcoxon, and his signed rank test has several applications. Its mechanism is best illustrated by an example.

EXAMPLE 6.5.1

The blood lead levels (in pg ml^{-1}) of seven children were found to be 104, 79, 98, 150, 87, 136 and 101. Could such data come from a population, assumed to be symmetrical, with a median/mean of 95 pg ml^{-1}?

On subtraction of the reference concentration (95) the data give values of

9, −16, 3, 55, −8, 41, 6

These values are first arranged in order of magnitude without regard to sign:

3, 6, 8, 9, 16, 41, 55

Their signs are then restored to them (in practice these last two steps can be combined):

3, 6, −8, 9, −16, 41, 55

The numbers are then **ranked**: in this process they keep their signs but are assigned numbers indicating their order (or rank):

1, 2, −3, 4, −5, 6, 7

The positive ranks add up to 20, and the negative ones to 8. The *lower* of these two figures (8) is taken as the test statistic. If the data came from a population with median 95 the sums of the negative and positive ranks would be expected to be approximately equal numerically; if the population median was very different from 95 the sums of the negative and positive ranks would be unequal. The probability of a particular sum occurring in practice is given by a set of tables (see Table A.11). In this test the null hypothesis is rejected if the experimental value is *less than or equal* to the tabulated value, i.e. the opposite of the situation encountered in most significance tests. In the present example examination of Table A.11 shows that, for $n = 7$, the test statistic must be less than or equal to 2 before the null hypothesis – that the data do come from a population of median (mean) 95 – can be rejected at a significance level of $P = 0.05$. In this example, the null hypothesis must be retained. As usual, a two-sided test is used though there may be occasional cases where a one-sided test is more appropriate.

An important advantage of the signed rank test is that it can also be used on paired data, because they can be transformed into the type of data given in the previous example. The signed rank method can thus be used as a non-parametric alternative to the paired *t*-test (Section 3.4).

EXAMPLE 6.5.2

The following table gives the percentage concentration of zinc, determined by two different methods, for each of eight samples of health food.

Sample	*EDTA titration*	*Atomic spectrometry*
1	7.2	7.6
2	6.1	6.8
3	5.2	4.6
4	5.9	5.7
5	9.0	9.7
6	8.5	8.7
7	6.6	7.0
8	4.4	4.7

Is there any evidence for a systematic difference between the results of the two methods? The approach to this type of problem is very simple. If there were no systematic difference between the two methods, then we would expect that the differences between the results for each sample, i.e. [(titration result) − (spectrometry result)], should be symmetrically distributed about zero. The signed differences are:

$$-0.4, \quad -0.7, \quad 0.6, \quad 0.2, \quad -0.7, \quad -0.2, \quad -0.4, \quad -0.3$$

Arranging these values in numerical order while retaining their signs, we have:

$$-0.2, \quad 0.2, \quad -0.3, \quad -0.4, \quad -0.4, \quad 0.6, \quad -0.7, \quad -0.7$$

The ranking of these results presents an obvious difficulty, that of *tied ranks*. There are two results with the numerical value 0.2, two with a numerical value of 0.4, and two with a numerical value of 0.7. How are the ranks to be calculated? This problem is resolved by giving the tied values average ranks, with appropriate signs. Thus the ranking for the present data is:

$$-1.5, \quad 1.5, \quad -3, \quad -4.5, \quad -4.5, \quad 6, \quad -7.5, \quad -7.5$$

In such cases, it is worth verifying that the ranking has been done correctly by calculating the sum of all the ranks without regard to sign. The sum for the numbers above is 36, which is the same as the sum of the first eight integers, and therefore correct. The sum of the positive ranks is 7.5, and the sum of the negative ranks is 28.5. The test statistic is thus 7.5. Inspection of Table A.11 shows that, for $n = 8$, the test statistic has to be ≤ 3 before the null hypothesis can be rejected at the level $P = 0.05$. In the present case, the null hypothesis must be retained – there is no evidence that the median (mean) of the difference is not zero, and hence no evidence for a systematic difference between the two analytical methods.

The signed rank test is seen from these examples to be a simple and valuable method. Its principal limitation is that it cannot be applied to very small sets of data: for a two-tailed test at the significance level $P = 0.05$, n must be at least 6.

6.6 Simple tests for two independent samples

The signed rank test described in the previous section is valuable for the study of single sets of measurements, and for paired sets that can readily be reduced to single sets. In many instances, however, it is necessary to compare two independent samples that cannot be reduced to a single set of data. Such samples may contain different numbers of measurements. Several non-parametric tests to tackle such problems have been devised. The simplest to understand and perform is the **Mann–Whitney *U*-test**, the operation of which is most easily demonstrated by an example.

EXAMPLE 6.6.1

A sample of photographic waste was analysed for silver by atomic absorption spectrometry, five successive measurements giving values of 9.8, 10.2, 10.7, 9.5, and 10.5 μg ml^{-1}. After chemical treatment, the waste was analysed again by the same procedure, five successive measurements giving values of 7.7, 9.7, 8.0, 9.9 and 9.0 μg ml^{-1}. Is there any evidence that the treatment produced a significant reduction in the levels of silver?

The Mann–Whitney procedure involves finding the number of results in one sample that exceeds each of the values in the other sample.

In the present case, we believe that the silver concentration of the treated solution should, if anything, be lower than that of the untreated solution (i.e. a one-sided test is appropriate). We thus expect to find that the number of cases in which a treated sample has a higher value than an untreated one should be small. Each of the values for the untreated sample is listed, and the number of instances where the values for the treated sample are greater are counted in each case.

Untreated sample	*Higher values in treated sample*	*Number of higher values*
9.8	9.9	1
10.2	–	0
10.7	–	0
9.5	9.7, 9.9	2
10.5	–	0

The total of the numbers in the third column, in this case 3, is the test statistic. Table A.12 is used for the Mann–Whitney *U*-test: again the critical values leading to the rejection of the null hypothesis are those which are *less than or equal* to the tabulated numbers. The table shows that for a one-sided test at $P = 0.05$, with five measurements of each sample, the test statistic must be ≤ 4 if the null hypothesis is to be rejected. In our example we can thus reject H_0: the treatment of the silver-containing material probably does reduce the level of the metal.

When, as in this example, the numbers of measurements are small the Mann–Whitney calculation can be done mentally, a great advantage. If ties (identical values) occur in the *U*-test, each tie is assigned a value of 0.5 in the count of *U*.

A further convenient method with some interesting features is **Tukey's quick test.** Its use can be shown using the same example.

Tukey's quick test involves counting the total number of measurements in the two independent samples that are not included in the overlap region of the two data sets.

EXAMPLE 6.6.2

Apply Tukey's quick test to the data of the previous example.

The test can be regarded as having two stages, though when only a few results are available, these two steps will doubtless be amalgamated into one rapid mental calculation. In the first step, the number of results in the second set of data that are *lower than all the values in the first set* are counted. If there are no such values, the test ends at once, and the null hypothesis of equal medians is accepted. In the present example, there are three such values, the readings 7.7, 8.0 and 9.0 being lower than the lowest value from the first set (9.5). The test thus continues to the second step, in which we count all the values in the first data set that are *higher than all the values in the second set.* Again, if there are no such values, the test ends and the null hypothesis is accepted. Here, there are again three such values, the readings 10.2, 10.5 and 10.7 exceeding the highest value in the second set (9.7). (This approach contrasts with that of the Mann–Whitney *U*-test, which identifies *high* values in the sample that might be expected to have the *lower* median.) Overall there are thus six values that are not within the range over which the two samples overlap. This total (often called T) is the test statistic. The most interesting and valuable aspect of Tukey's quick test is that statistical tables are not normally needed to interpret this result. Provided that the number of readings in each sample does not exceed about 20, and that the two sample sizes are not greatly different (conditions that would probably be valid in most analytical laboratory experiments), the critical values of T for a particular level of significance are *independent of sample size*. For a one-sided test the null hypothesis may be rejected if $T \geq 6$ (for $P = 0.05$), ≥ 7 ($P = 0.025$), ≥ 10 ($P = 0.005$), and ≥ 14 ($P = 0.0005$). (For a two-tailed test the critical T values at $P = 0.05$, 0.025, 0.005 and 0.0005 are 7, 8, 11 and 15 respectively.) In the present example, therefore, the experimental T value is big enough to be significant at $P = 0.05$ in a one-sided test. We can thus reject the null hypothesis and report that the treatment does reduce the silver content of the photographic waste significantly, a result in accord with that of the Mann–Whitney *U*-test.

If ties occur in Tukey's quick test (i.e. if one of the values in the hypothetically higher sample is equal to the highest value in the other sample, or if one of the values in the 'lower' sample is equal to the lowest value in the 'higher' sample) then each tie counts 0.5 towards the value of T.

A test which is distantly related to the Mann–Whitney method has been developed by Siegel and Tukey to compare the spread of two sets of results, and thus offer a genuinely non-parametric alternative to the *F*-test (see Section 3.6). The data from the two sets of measurements are first pooled and arranged in numerical order, but with one set of results distinguished by underlining. Then they are ranked in an ingenious way: the lowest measurement is ranked one, the highest measurement ranked two, the highest but one measurement ranked three, the lowest but one ranked four, the lowest but two ranked five, and so on. (If the total number of measurements is odd, the central measurement is ignored.) This *paired alternate ranking* produces a situation in which the low and high results receive low ranks, and the central results receive high ranks. If one data set has a significantly wider spread than the other, its sum of ranks should thus be much lower, while if the dispersion of the two sets of results is similar, their rank sums will be similar. Application of this method to the data from Example 6.6.1 gives the following rankings:

Data	7.7	8.0	9.0	9.5	9.7	9.8	9.9	10.2	10.5	10.7
Ranks	1	4	5	8	9	10	7	6	3	2

Two *rank sums* are then calculated. The sum of the underlined ranks (treated silver-containing samples) is 26, and the rank sum for the untreated samples is 29. In this example the sample sizes for the two sets of measurements are equal, but this will not always be the case. Allowance is made for this by subtracting from the rank sums the number $n_i(n_i + 1)/2$, where the n_i values are the sample sizes. In our example $n_i = 5$ in each case, so 15 must be subtracted from each rank sum. The lower of the two results is the one used in the test, and the critical values are the same as those used in the Mann–Whitney test (Table A.12). The test statistic obtained in this example is $(26 - 15) = 11$, much higher than the critical value of 2 (for a two-tailed test at $P = 0.05$). The null hypothesis, in this case that the spread of the results is similar for the two sets of data, is thus retained.

> The Siegel–Tukey test pools the two data samples with identification, ranks them, applies paired alternate ranking to generate rank sums, and allows for the sample sizes, to provide a test statistic that can be evaluated using the same tables as for the Mann–Whitney *U*-test.

A little thought will show that the validity of this useful test will be reduced if the average values for the two sets of data are substantially different. In the extreme case where all the measurements in one sample are lower than all the measurements in the other sample, the rank sums will always be as similar as possible, whatever the spread of the two samples. If it is feared that this effect is appreciable, it is permissible to estimate the means of the two samples, and add the difference between the means to each of the measurements of the lower set. This will remove any effect due to the different means, while preserving the dispersion of the sample. An example of the application of this test is provided at the end of the chapter.

6.7 Non-parametric tests for more than two samples

The previous section described tests in which two statistical samples were compared with each other. Non-parametric methods are not, however, limited to two sets of data: several methods which compare three or more samples are available. Before two of these tests are outlined, it is important to mention one pitfall to be avoided in all multi-sample comparisons. When (for example) three sets of measurements are examined to see whether or not their medians are similar, there is a great temptation to compare only the two samples with the highest and lowest medians. This simplistic approach can give misleading results. When several samples are taken from the *same parent population*, there are cases where the highest and lowest medians, considered in isolation, appear to be significantly different. This is because, as the number of samples increases, the difference between the highest and the lowest medians will tend to increase. The correct approach is to perform first a test that considers *all the samples together*: if it shows that they might not all come from the same population, then separate tests may be performed to try to identify where the significant differences occur. Here we describe in outline the principles of two non-parametric tests for three or more sets of data: the reader seeking further detail is recommended to consult the books given in the bibliography.

The Kruskal–Wallis test is applied to the comparison of the medians of three or more unmatched samples. (An extension of the silver analysis described in the previous section, with three samples of photographic waste, one untreated and the other two treated by different methods, would provide an instance where the test would be useful.) The results from the three (or more) samples are pooled and arranged in rank order. The rank totals for the data from the different samples are determined: tied ranks are averaged, as shown above, though a special correction procedure is advisable if there are numerous ties. If each sample has the same number of measurements (this is not a requirement of the test), and if the samples have similar medians, then the rank totals for each sample should be similar, and the sum of their squares should be a minimum. For example, if we have three samples, each with five measurements, the rankings will range from 1 to 15 and the sum of all the ranks will be 120. Suppose that the three medians are very similar, and that the rank totals for each sample are thus equal, each being 40. The sum of the squares of these totals will thus be $40^2 + 40^2 + 40^2 = 4800$. If the medians are significantly different, then the rank totals will also be different from one another – say 20, 40, and 60. The sum of the squares of such totals will always be larger than 4800 ($20^2 + 40^2 + 60^2 = 5600$).

The probability of obtaining any particular sum of squares can be determined by using the **chi-squared** statistic (see Chapter 3). If the samples are referred to as A, B, C, etc. (k samples in all), with numbers of measurements n_A, n_B, n_C, etc. and rank totals R_A, R_B, R_C, etc., then the value of X^2 is given by:

$$X^2 = \frac{12}{N^2 + N}\left\{\frac{R_A^2}{n_A} + \frac{R_B^2}{n_B} + \frac{R_C^2}{n_C} + \dots\right\} - 3(N+1) \qquad (6.2)$$

where $N = n_A + n_B + n_C$, etc. This X^2 value is compared as usual with tabulated values. The latter are identical to the usual values when the total number of measurements is more than ca. 15, but special tables are used for smaller numbers of measurements. The number of degrees of freedom is $k - 1$. Experimental values of X^2 that exceed the tabulated values allow the null hypothesis (that the medians of the samples are not significantly different) to be rejected. As already noted, in the latter situation further tests can be performed on individual pairs of samples: again, texts listed in the bibliography provide more details.

We have already seen (Sections 3.4 and 6.3) that when *paired* results are compared, special statistical tests can be used. These tests use the principle that, when two experimental methods that do not differ significantly are applied to the same chemical samples, the differences between the matched pairs of results should be close to zero. This principle can be extended to three or more matched sets of results by using a non-parametric test devised in 1937 by Friedman. In analytical chemistry, the main application of **Friedman's test** is in the comparison of three (or more) experimental methods applied to the same chemical samples. The test again uses the X^2 statistic, in this case to assess the differences that occur between the total rank values for the different methods. The following example illustrates the simplicity of the approach.

EXAMPLE 6.7.1

The levels of a pesticide in four plant extracts were determined by (A) high-performance liquid chromatography, (B) gas–liquid chromatography, and (C) radioimmunoassay. The following results (all in ng ml^{-1}) were obtained:

Sample	*Method*		
	A	*B*	*C*
1	4.7	5.8	5.7
2	7.7	7.7	8.5
3	9.0	9.9	9.5
4	2.3	2.0	2.9

Do the three methods give values for the pesticide levels that differ significantly?

This problem is solved by replacing the values in the table by ranks. In each row the method with the lowest result is ranked 1, and that with the highest result is ranked 3:

Sample	*Method*		
	A	*B*	*C*
1	1	3	2
2	1.5	1.5	3
3	1	3	2
4	2	1	3

The use of an average value is necessary in the case of tied ranks in sample 2 (see Section 6.5). The sums of the ranks for the three methods A, B and C are 5.5, 8.5 and 10 respectively. These sums should total $nk(k+1)/2$, where k is the number of methods (three here) and n the number of samples (four here). The rank sums are squared, yielding 30.25, 72.25 and 100 respectively, and these squares are added to give the statistic R, which here is 202.5. The experimental value of X^2 is then calculated from:

$$X^2 = \frac{12R}{nk(k+1)} - 3n(k+1) \tag{6.3}$$

which gives a result of 2.625. At the level $P = 0.05$, and with $k = 3$, the critical values of X^2 are 6.0, 6.5, 6.4, 7.0, 7.l and 6.2 for $n = 3$, 4, 5, 6, 7 and 8 respectively. (More extensive data are given in many sets of statistical tables, and when $k > 7$ the usual X^2 tables can be used at $k - 1$ degrees of freedom.) In this instance, the experimental value of X^2 is much less than the critical value, and we must retain the null hypothesis: the three methods give results that do not differ significantly.

The Friedman test could alternatively be used in the reverse form: assuming that the three analytical methods give indistinguishable results, the same procedure could be used to test differences between the four plant extracts. In this case k and n are 4 and 3 respectively, and the reader may care to verify that R is 270 and that the resulting X^2 value is 9.0. This is higher than the critical value for $P = 0.05$, $n = 3$, $k = 4$, which is 7.4. So in this second application of the test we can reject the null hypothesis, and state that the four samples do differ in their pesticide levels. Further tests, which would allow selected comparisons between pairs of samples, are then available.

Friedman's test is evidently much simpler to perform in practice than the ANOVA method (Sections 3.8–3.10), though it does not have the latter's ability to study interaction effects (see Chapter 7).

6.8 Rank correlation

Ranking methods can also be applied to correlation problems. The Spearman rank correlation coefficient method to be described in this section is the oldest application of ranking methods in statistics, dating from 1904. Like other ranking methods, it is of particular advantage when one or both of the sets of observations under investigation can be expressed only in terms of a rank order rather than in quantitative units. Thus, in the following example, the possible correlation between the sulphur dioxide concentrations in a series of table wines and their taste quality is investigated. The taste quality of a wine is not easily expressed in quantitative terms, but it is relatively simple for a panel of wine-tasters to rank the wines in order of preference. Examples of other attributes that are easily ranked, but not easily quantified, include the condition of experimental animals, the quality of laboratory accommodation, and the efficiency of laboratory staff. It should also be remembered that if either or both the sets of data under study should happen to be quantitative, then (in contrast to the methods described in Chapter 5) there is no need for them to be normally distributed. Like other non-parametric statistics, the Spearman rank correlation coefficient, r_s, is easy to determine and interpret. This is shown in the following example.

EXAMPLE 6.8.1

Seven different table wines are ranked in order of preference by a panel of experts. The best wine is ranked 1, the next best 2, and so on. The sulphur dioxide content (in parts per million) of each wine is then determined by flow injection analysis with colorimetric detection. Use the following results to determine whether there is any relationship between perceived wine quality and sulphur dioxide content.

Wine	A	B	C	D	E	F	G
Taste ranking	1	2	3	4	5	6	7
SO_2 content	0.9	1.8	1.7	2.9	3.5	3.3	4.7

The first step in the calculation is to convert the sulphur dioxide concentrations from absolute values into ranks (tied ranks are averaged as described in previous sections):

Wine	A	B	C	D	E	F	G
Taste ranking	1	2	3	4	5	6	7
SO_2 content	1	3	2	4	6	5	7

The differences, d_i, between the two ranks are then calculated. They are 0, −1, 1, 0, −1, 1, 0. The correlation coefficient, r_s, is then given by:

$$r_s = 1 - \frac{6\sum_i d_i^2}{n(n^2-1)} \tag{6.4}$$

In this example, r_s is $1-(24/336)$, i.e. 0.929. Theory shows that, like the product-moment correlation coefficient, r_s can vary between −1 and +1. When $n = 7$, r_s must exceed 0.786 if the null hypothesis of no correlation is to be rejected at the significance level $P = 0.05$ (Table A.13). Here, we can conclude that there is a correlation between the sulphur dioxide content of the wines and their perceived quality. Bearing in mind the way the rankings were defined, there is strong evidence that higher sulphur dioxide levels produce less palatable wines!

Another rank correlation coefficient, due to Kendall, was introduced in 1938. It claims to have some theoretical advantages over the Spearman method, but is harder to calculate (especially when tied ranks occur) and is not so frequently used.

6.9 Non-parametric regression methods

In the detailed discussion of linear regression methods in the previous chapter, the assumption of normally distributed y-direction errors was emphasized, and the complexity of some of the calculation methods was apparent. This complexity is largely overcome by using calculators or computers, and there are also some rapid approximate methods for fitting straight lines to experimental data (see Bibliography). There is still an interest in non-parametric approaches to fitting a straight line to a set of points. Of the several methods available, perhaps the simplest is **Theil's 'incomplete' method,** so called to distinguish it from another more complex procedure developed by the same author (the 'complete' method).

> Theil's method determines the slope of a regression line as the median of the slopes calculated from selected pairs of points: the intercept of the line is the median of the intercept values calculated from the slope and the co-ordinates of the individual points.

The method assumes that a series of points (x_1, y_1), (x_2, y_2), etc. is fitted by a line of the form $y = a + bx$. The first step in the calculation involves ranking the points in order of increasing x. If the number of points, x, is odd, the middle point, i.e. the median value of x, is deleted: the calculation always requires an even number of points. For any pair of points (x_i, y_i), (x_j, y_j) where $x_j > x_i$, the slope, b_{ij}, of the line joining the points can be calculated from:

$$b_{ij} = \frac{(y_j - y_i)}{(x_j - x_i)} \tag{6.5}$$

Slopes b_{ij} are calculated for the pair of points (x_1, y_1) and the point immediately after the median x-value, for (x_2, y_2) and the second point after the median x-value, and so on until the slope is calculated for the line joining the point immediately before the median x with the last point. Thus, if the original data contained 11 points, five slopes would be estimated (the median point having been omitted). For eight original points there would be four slope estimates, and so on. These slope estimates are arranged in ascending order and their median is the estimated slope of the straight line. With this value of b, values a_i for the intercept are estimated for each point with the aid of the equation $y = a + bx$. Again the estimates of a are arranged in ascending order and the median value is chosen as the best estimate of the intercept of the line. The method is illustrated by the following example.

EXAMPLE 6.9.1

The following results were obtained in a calibration experiment for the absorptiometric determination of a metal chelate complex:

Concentration, μg ml^{-1}	0	10	20	30	40	50	60	70
Absorbance	0.04	0.23	0.39	0.59	0.84	0.86	1.24	1.42

Use Theil's method to estimate the slope and the intercept of the best straight line through these points.

In this case the calculation is simplified by the occurrence of an even number of observations, and by the fact that the *x*-values (i.e. the concentrations) occur at regular intervals and are already in ranking order. We thus calculate slope estimates from four pairs of points:

$$b_{15} = (0.84 - 0.04)/40 = 0.0200$$
$$b_{26} = (0.86 - 0.23)/40 = 0.0158$$
$$b_{37} = (1.24 - 0.39)/40 = 0.0212$$
$$b_{48} = (1.42 - 0.59)/40 = 0.0208$$

We now arrange these slope estimates in order, obtaining 0.0158, 0.0200, 0.0208, 0.0212. The median estimate of the slope is thus the average of 0.0200 and 0.0208, i.e. 0.0204. We now use this value of b to estimate the intercept, a. The eight individual a_i values are:

$$a_1 = 0.04 - (0.0204 \times 0) = +0.040$$
$$a_2 = 0.23 - (0.0204 \times 10) = +0.026$$
$$a_3 = 0.39 - (0.0204 \times 20) = -0.018$$
$$a_4 = 0.59 - (0.0204 \times 30) = -0.022$$
$$a_5 = 0.84 - (0.0204 \times 40) = +0.024$$
$$a_6 = 0.86 - (0.0204 \times 50) = -0.160$$
$$a_7 = 1.24 - (0.0204 \times 60) = +0.016$$
$$a_8 = 1.42 - (0.0204 \times 70) = -0.008$$

Arranging these intercept estimates in order, we have −0.160, −0.022, −0.018, −0.008, +0.016, +0.024, +0.026, +0.040. The median estimate is +0.004. So the best straight line is given by $y = 0.0204x + 0.004$. The 'least squares' line, calculated by the methods of Chapter 5, is $y = 0.0195x + 0.019$. Figure 6.4 shows that the two lines are very similar when plotted. However, the Theil method has three distinct advantages: it does not assume that all the errors are in the *y*-direction; it does not assume that either the *x*- or *y*-direction errors are normally distributed; and it is not affected by the presence of outlying results. This last point is clearly illustrated by the point (50, 0.86) in the present example. It has every appearance of being an outlier, but its value does not affect the Theil calculation at all, since neither b_{26} nor a_6 directly affects the median estimates of the slope and intercept respectively. In the least squares calculation, however, this outlying point carries as much weight as the other points. This is reflected in the calculated results; the least squares line passes closer to the outlier than the non-parametric line does.

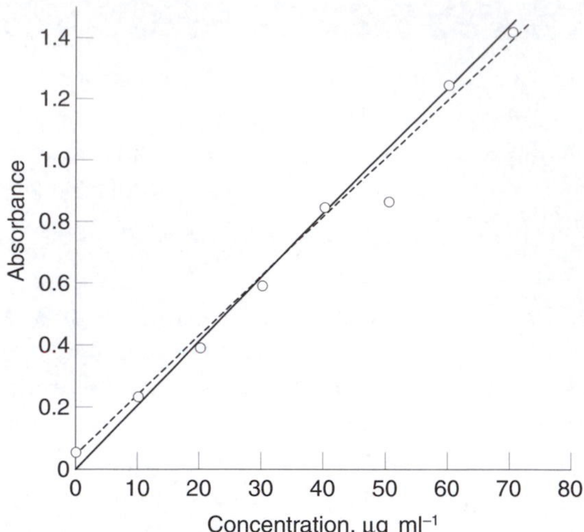

Figure 6.4 Straight-line calibration graph calculated by Theil's method (———), and by the least squares method of Chapter 5 (— — —).

Unlike most non-parametric methods, Theil's method involves fairly tedious calculations, so a computer program such as a spreadsheet macro is necessary in practice. It should be noted that non-parametric methods for fitting curves are also available, but these are beyond the scope of the present book.

6.10 Robust methods

At the beginning of this chapter, it was noted that there is growing evidence for the occurrence in the experimental sciences of *heavy-tailed* error distributions. These can be regarded as normal (Gaussian) distributions with the addition of outliers arising from gross errors, or as the result of the superposition of several normal distributions with similar means but different variances. In either case, and in other instances where the departure from a normal distribution is not great, it seems to be a waste of information to use non-parametric methods, which make no assumptions at all about the underlying error distribution. A better approach would be to develop methods which do not entirely exclude suspicious results, but which *reduce the weight* given to such data. This is the philosophy underlying the robust methods to be summarized in this section and the next: such methods can be applied to repeated measurements and also to calibration/regression data. Many robust methods have been developed, so only a brief survey of this rapidly developing field is possible here: the reader is referred to the Bibliography for sources of further material.

An obvious problem occurs in virtually all these methods. If we are to downgrade the significance of some of our measurements, one or more criteria are needed on which to base such decisions, but we cannot use such criteria unless we initially consider all the data. This problem is solved by using *iterative* methods: we estimate or guess a starting value or values for some property of our data, use such initial estimates with our weighting criteria to arrive at a second estimate, then we re-apply our criteria, etc. Such methods are only practicable if a computer is available, though it must be stressed that many otherwise excellent suites of statistics software do not yet include programs for robust methods.

There are some very simple robust methods that do not require such iterations, because they arbitrarily eliminate, rather than downweight, a proportion of the data. For example, the **trimmed mean** for any set of data is found by omitting r observations at the top and at the bottom of the range of measurements. This principle can be applied to the set of data in one of the examples in Section 3.7. This example considered seven replicate measurements of nitrite ion in river water (mg l^{-1}):

0.380, 0.400, 0.401, 0.403, 0.410, 0.411, 0.413

The data have been arranged in numerical order for convenience: this emphasizes that the obvious question is whether or not the measurement 0.380 is an outlier. If the number 0.380 is retained, the mean of the seven measurements is 0.4026, and their standard deviation is 0.0112. If, as Dixon's test (Section 3.7) suggests, it is permissible to reject the result 0.380 (at $P = 0.05$), the mean and standard deviation then become 0.4063 and 0.0056 respectively. This confirms, as noted in Section 3.3, that the mean and (especially) the standard deviation are vulnerable to the occurrence of outliers. Now suppose that we omit the smallest (0.380) and the largest (0.413) of the above measurements, and recalculate the mean. This produces a number technically known as the 14.28% trimmed mean, the percentage being calculated as $100r/n$ where r top and bottom measurements have been omitted from n results. This trimmed mean is 0.4050, clearly closer to the second of the two means calculated above, i.e. the mean determined after rejection of the outlier. The robustness of this trimmed mean is obvious – it would have been the same, *whatever* the values taken by the smallest and largest results. But this also illustrates the crudity of the trimmed mean method. Why should we omit the value 0.413, except for reasons of symmetry? Is it acceptable to calculate a statistic that *completely* ignores the suspect result(s)? What percentage of the data should be removed by trimming? (In practice 10–25% trimming is common.) Procedures that are superior to crude trimming methods are discussed below.

A simple robust estimate of the standard deviation is provided by the interquartile range (IQR, see Section 6.2). For a normal error distribution, the IQR is ca. 1.35σ. This relationship supplies a standard deviation estimate that is not affected by any value taken by the largest or smallest measurements. Unfortunately, the IQR is not a very meaningful concept for very small data sets. Moreover, and somewhat surprisingly, there are several different conventions for its calculation. For large samples the convention chosen makes little difference, but for small samples the differences in the calculated IQR values are large, so the IQR has little application in analytical chemistry.

A more logical approach to robust estimation can be based on the concept of a **distance function**. Suppose we have a series of n results $x_1 \ldots x_n$, and we wish to estimate μ, the mean of the 'reliable' results. Normally our estimate of μ, given here the symbol $\hat{\mu}$, is obtained by minimizing the sum of squares (SS) $\sum_i (x_i - \mu)^2$. (This sum of *squared* terms is the source of the sensitivity of the mean to large errors.) The expression $(x - \mu)^2$ is referred to as a distance function, since it measures the distance of a point from μ. A more useful distance function in the present context is $|x - \mu|$. A widely used method to test measurements for downweighting data is to compare $|x - \mu|$ with $c\sigma$, where c is usually taken to be 1.5 and σ^2 is a robust estimate of variance. We consider first the estimation of σ^2, and then discuss the downweighting procedure.

The robust variance estimate can be derived from a statistic related to the unfortunately abbreviated **median absolute deviation** (MAD!), which is calculated from

$$\text{MAD} = \text{median}[|x_i - \text{median}(x_i)|] \tag{6.6}$$

The MAD is an extremely useful statistic: one rough method for evaluating outliers (x_0) is to reject them if $[|x_0 - \text{median}(x_i)|]/\text{MAD} > 5$. It can be shown that MAD/0.6745 is a useful robust estimate of σ (called $\hat{\sigma}$) which can be used unchanged during the iterative estimates of $\hat{\mu}$.

EXAMPLE 6.10.1

These techniques can be applied to the measurements discussed above (0.380, 0.400, 0.401, 0.403, 0.410, 0.411, 0.413). First it is necessary to calculate the MAD. The median of these numbers is 0.403 (i.e. the fourth of the seven ordered values), so the individual deviations (without regard to their signs) are 0.023, 0.003, 0.002, 0, 0.007, 0.008, and 0.010. Rewriting these in numerical order we have: 0, 0.002, 0.003, 0.007, 0.008, 0.010, and 0.023. The MAD is the median of these seven numbers, i.e. 0.007, so $\hat{\sigma} = \text{MAD}/0.6745 = 0.007/0.6745 = 0.0104$, and $1.5\hat{\sigma}$ is 0.0156.

We are now in a position to begin iterative estimates of $\hat{\mu}$. This process is begun by taking any reasonable estimate for $\hat{\mu}$ and calculating $|x_i - \hat{\mu}|$ values for each measurement. In this example, suppose the initial $\hat{\mu}$-value is the median, 0.403. As we have seen, the individual deviations from this value are (in numerical order, but neglecting their signs) 0, 0.002, 0.003, 0.007, 0.008, 0.010, and 0.023. In the first iteration for $\hat{\mu}$ the original measurements are retained if these deviations from the median are ≤ 0.0156. This applies to all the deviations listed except the last. In the event that the deviation is >0.0156, the original value in question is *changed* to become $\hat{\mu} - c\hat{\sigma}$ or $\hat{\mu} + c\hat{\sigma}$, depending on whether it was originally below or above the median respectively. In the present example, the value 0.380, which gives rise to the large deviation of 0.023, has to be changed to $\hat{\mu} - c\hat{\sigma}$, i.e. $0.403 - 0.0156 = 0.3874$.

There is thus now a new data set, with the measurement 0.380 in the original set having been replaced by 0.3874. This new set of numbers is called a set of *pseudo-values* ($\tilde{x}_i$), and the calculation is repeated using this new set. The first step is to calculate the mean of the new values (note that although the initial value of $\hat{\mu}$ can be based on the mean or the median or any other sensible estimate, subsequent steps in the iteration always use the mean): this gives the result 0.4036. The individual deviations from this new estimate of $\hat{\mu}$ are, in numerical order and without signs, 0.0006, 0.0026, 0.0036, 0.0064, 0.0074, 0.0094, and 0.0162. As expected (since only one measurement was suspicious in the first place) only the last of these deviations exceeds 0.0156, which means that again the measurement in question is changed, from 0.3874 to (0.4036 – 0.0156) = 0.3880. The next set of seven values is thus the same as the previous set, except that the value 0.3874 is replaced by 0.3880. The new mean ($\hat{\mu}$-value) is thus 0.4037. This is so close to the previous value that it is clearly unnecessary to undertake any further iterations: we conclude that a robust estimate of $\hat{\mu}$ is 0.4037, say 0.404. This example is typical in that there is a very rapid convergence of the iterated values of $\hat{\mu}$.

This calculation deserves several comments. The first is that, like so many iterative procedures, it is much more tedious to describe and explain than it is to perform! The second point to note is that in this example we have estimated μ by making some (robust and reasonable) assumptions about $\hat{\sigma}$. There are also methods where the opposite is true, i.e. a robust estimate of precision is obtained if a mean value is known, and yet more methods in which both robust estimates are calculated iteratively and side by side. Lastly it is worth re-emphasizing that these robust methods do not have the worries and ambiguities of outlier tests. In the example just examined, the Dixon test (Section 3.7) suggested that the value 0.380 could be rejected as an outlier ($P = 0.05$), but the simple MAD-based test (see above) suggests that it should not, as $[|x_0 - \text{median}(x_i)|]/\text{MAD} = [|0.380 - 0.403|]/0.007 = 3.3$, well below the (rough) critical value of 5. Such concerns and contradictions disappear in robust statistics, where the outliers are neither wholly rejected nor accepted unchanged, but accepted in a changed or down-weighted form.

Another robust approach to repeated measurement statistics (and to regression) is provided by Winsorization. This can be regarded as a variant of the method described above. The measurements yielding the largest positive or negative deviations from the median (or, in regression, the largest residuals) are reduced in importance by moving them so that their deviations/residuals are equal to the next largest (or perhaps the third largest) positive or negative values respectively. The arbitrariness of the procedure is less than in trimming methods, as the change in the deviation/residual value for any 'sensible' point is small. Given the availability of the required software, these and other robust regression techniques are sure to find increasing use in analytical chemistry in future. One area where their use is already recommended is in inter-laboratory comparisons (see Chapter 4).

6.11 Robust regression methods

The problems caused by possible outliers in regression calculations have been outlined in Sections 5.13 and 6.9, where rejection using a specified criterion and non-parametric approaches respectively have been discussed. It is clear that robust approaches will be of value in regression statistics as well as in the statistics of repeated measurements, and there has indeed been a rapid growth of interest in robust regression methods amongst analytical scientists. A summary of two of the many approaches developed must suffice.

In Section 6.9 it was noted that a single suspect measurement has a considerable effect on the *a* and *b* values calculated for a straight line by the normal 'least squares' method, which seeks to minimize the sum of the squares of the *y*-residuals. This is because, just as in the nitrite determination example given above, the use of squared terms causes such suspect data points to have a big influence on the sum of squares. A clear and obvious alternative is to seek to minimize the *median* of the squared residuals, which will be much less affected by large residuals. This **least median of squares** (LMS) method is very robust: its *breakdown point,* i.e. the proportion of outliers amongst the data that it can tolerate, is 50%, the theoretical maximum value. (If the proportion of 'suspect' results exceeds 50% it clearly becomes impossible to distinguish them from the 'reliable' results.) Simulations using data sets with deliberately included outliers show that this is a much better performance than that obtained with the Theil method. The LMS method also works well in cases such as that discussed in Section 5.11, where we wish to characterize the straight-line portion of a set of data which is linear near to the origin, but non-linear at higher *x* and *y* values. Its disadvantage is that it involves an iterative calculation which converges rather slowly: that is, many iterations are often required before the estimated *a* and *b* values become more or less constant.

Other robust regression methods are being increasingly used. The **iteratively re-weighted least squares** method begins with a straightforward least-squares estimate of the parameters of a line. The resulting residuals are then given different weights, usually via a **biweight** approach. Points with very large residuals (e.g. at least six times greater than the median residual value) are rejected, i.e. given zero weight, while points with smaller residuals are given weights which increase as the residuals themselves get smaller. A *weighted* least squares calculation (Section 5.10) is then applied to the new data set, and these steps are repeated until the values for *a* and *b* converge to stable levels. In this method convergence is usually fairly rapid.

6.12 The Kolmogorov test for goodness of fit

In Chapter 3 the common statistical problem of 'goodness of fit' was discussed. This problem arises when it is required to test whether a sample of observations might come from a particular distribution, such as a normal distribution. The chi-squared test is very suitable for this purpose when the data are presented as frequencies, though the test is not normally used for fewer than about 50 observations, and is difficult to use with continuous data. In this section, the Kolmogorov method, which is well suited to testing goodness of fit with continuous data, is described. Extensions of the method, which will not be described in detail, allow it to be applied to the comparison of two samples. These modified methods were first discovered by Smirnov, and the series of tests is often known as the **Kolmogorov–Smirnov** method.

The principle of the Kolmogorov approach is very simple. It involves comparing the cumulative frequency curve of the data to be tested with the cumulative frequency curve of the hypothesized distribution. The concept of the cumulative frequency curve, and its application in conjunction with normal probability paper, was discussed in Chapter 3. When the hypothetical and experimental curves have been drawn, the test statistic is obtained by finding the maximum vertical difference between them, and comparing this value in the usual way with a set of tabulated values. If the experimental data depart substantially from the expected distribution, the two curves will be widely separated over part of the cumulative frequency diagram. If, however, the data are closely in accord with the expected distribution, the two curves will never be very far apart. In practice, the Kolmogorov method has two common applications – testing for randomness, and testing for normality of distribution – and a simple example of the latter application will illustrate the operation of the procedure.

When the Kolmogorov method is used to test whether a distribution is normal, we must first transform the original data, which might have any values for their mean and standard deviation, into the **standard normal variable**, z (see Section 2.2). This is done by using the equation:

$$z = \frac{x - \mu}{\sigma} \tag{6.7}$$

where the terms have their usual meanings. The cumulative distribution function is given in Table A.1. Equation (6.7) can be used in two ways. In some cases it may be required to test whether a set of data could come from a *particular* normal distribution, of *given* mean and standard deviation. In such a case, the experimental data are transformed directly by using equation (6.7), and the Kolmogorov test is performed. More often, it will simply be required to test whether the data might come from any normal distribution. In this instance, the mean and the standard deviation are estimated first, by the methods of Chapter 2; the data are next transformed by using equation (6.7); then the Kolmogorov method is applied but using different critical values. Both types of test are illustrated in the following example.

EXAMPLE 6.12.1

Eight titrations were performed, with the results 25.13, 25.02, 25.11, 25.07, 25.03, 24.97, 25.14, and 25.09 ml. Could such results have come from (a) a normal population with mean 25.00 ml and standard deviation 0.05 ml, and (b) from any other normal population?

(a) Here, the first step is to transform the x-values into z-values by using the relationship $z = (x - 25.00)/0.05$, obtained from equation (6.7). The eight results are thus transformed into 2.6, 0.4, 2.2, 1.4, 0.6, −0.6, 2.8, and 1.8. These z-values are arranged in order and plotted as a cumulative distribution function with a step height of 0.125 (i.e. 1/8). (Note that this is not quite the same approach as that used in Section 3.12.) Comparison with the hypothetical function for z (Table A.1) indicates (Figure 6.5) that the maximum difference is 0.545 at a z-value just below 1.4. To test this value, Table A.14 is used. The table shows that, for $n = 8$ and $P = 0.05$, the critical value for a specified distribution is 0.454, so the null hypothesis can be rejected – the titration values probably do not come from a normal population with mean 25.00 ml and standard deviation 0.05 ml.

(b) In this case, we estimate the mean and the standard deviation [with the aid of equations (2.1) and (2.2)] as 25.07 and 0.059 ml respectively, the latter result being correct to two significant figures. The z-values are now given by $z = (x - 25.07)/0.059$, i.e. by 1.02, −0.85, 0.68, 0, −0.68, −1.69, 1.19, 0.34. The cumulative frequency diagram for these values differs from the hypothetical curve by 0.125 at most (at several points). This difference is much smaller than the critical value of 0.288. We can thus accept the null hypothesis that the data come from a normal population with mean 25.07 and standard deviation 0.059.

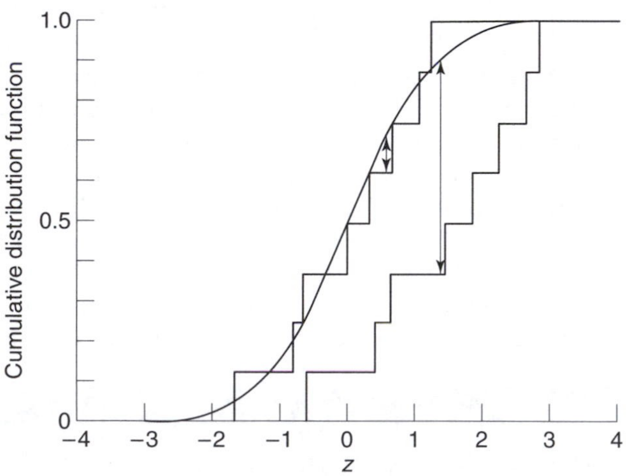

Figure 6.5 Kolmogorov's method used to test for the normal distribution. Maximum differences between the theoretical cumulative frequency curve and the two tested distributions are shown by the arrows (⟷).

6.13 Conclusions

The robust and non-parametric tests described in this chapter are only a small fraction of the total available number of such methods. The examples given exemplify their strengths and weaknesses. In many cases their speed and convenience give them a distinct advantage over conventional methods, and the non-parametric tests do not involve the assumption of a normal distribution. They are ideally suited to the preliminary examination of small numbers of measurements, and to quick calculations made – often without the need for tables – while the analyst is at the bench or on the shop floor. They can also be used when three or more samples are studied (Section 6.7). The **power** (i.e. the probability that a false null hypothesis is rejected: see Section 3.13) of a non-parametric method may be less than that of the corresponding parametric test, but the difference is only rarely serious. For example, many comparisons have been made of the powers in different conditions (i.e. different population distributions and sample sizes) of the Mann–Whitney U-test and the t-test. The former performs very well in almost all circumstances and is only marginally less powerful than the t-test even when the data come from a normally distributed population. Many suites of programs for personal computers now include several non-parametric tests. Such programs allow a particular set of data to be evaluated rapidly by two or more methods, and seem certain to enhance rather than reduce interest in these very convenient methods.

Robust methods are not normally so easy to use, in view of the need for iterative calculations, but they represent the best way of tackling one of the most common and difficult problems for practising analysts, the occurrence of suspicious or outlying results.

Overall a great variety of significance tests – parametric, non-parametric, and robust - are available, and often the most difficult task in practice is to decide which method is best suited to a particular problem. The diagram in Appendix 1 is designed to make such choices easier, though inevitably it cannot cover all possible practical situations.

Bibliography

Chatfield, C. 1988. *Problem Solving: A Statistician's Guide.* Chapman and Hall, London. (A thoroughly practical and useful book containing much valuable advice: lengthy appendices summarize the background theory.)

Conover, W. J. 1971. *Practical Non-parametric Statistics*. Wiley, New York. (Probably the best known general text on non-parametric methods.)

Daniel, W. 1978. *Applied Nonparametric Statistics*. Houghton Mifflin, Boston. (A very comprehensive text, covering a wide range of non-parametric methods in considerable detail; many examples.)

Rousseeuw, P. J. and Leroy, A. M. 1987. *Robust Regression and Outlier Detection.* John Wiley, New York. (An excellent and remarkably readable book with many illustrative examples.)

Sprent, P. 1981. *Quick Statistics.* Penguin Books, London. (An excellent introduction to non-parametric methods, with clear, non-mathematical explanations, and many examples and exercises.)

Sprent, P. 1989. *Applied Nonparametric Statistical Methods.* Chapman and Hall, London. (Covers a wide range of significance tests in a practical way, and with a good discussion of robust techniques.)

Velleman, P. F. and Hoaglin, D. C. 1991. *Applications, Basics and Computing of Exploratory Data Analysis.* Duxbury Press, Boston. (An excellent introduction to IDA/EDA: computer programs in BASIC and FORTRAN are provided.)

Exercises

1. A titration was performed four times, with the results: 9.84, 9.91, 9.89, 10.20 ml. Calculate and comment on the median and the mean of these results.

2. The level of sulphur in batches of an aircraft fuel is claimed by the manufacturer to be symmetrically distributed with a median value of 0.10%. Successive batches are found to have sulphur concentrations of 0.09, 0.12, 0.10, 0.11, 0.08, 0.17, 0.12. 0.14 and 0.11%. Use the sign test and the signed rank test to check the manufacturer's claim.

3. The concentrations (g 100 ml^{-1}) of immunoglobulin G in the blood sera of ten donors are measured by radial immunodiffusion (RID) and by electro-immunodiffusion (EID), with the following results:

Donor	1	2	3	4	5	6	7	8	9	10
RID result	1.3	1.5	0.7	0.9	1.0	1.1	0.8	1.8	0.4	1.3
EID result	1.1	1.6	0.5	0.8	0.8	1.0	0.7	1.4	0.4	0.9

   Are the results of the two methods significantly different?

4. Ten carbon rods used successively in an electrothermal atomic-absorption spectrometer were found to last for 24, 26, 30, 21, 19, 17, 23, 22, 25 and 25 samples. Test the randomness of these rod lifetimes.

5. After each drinking three pints of beer, five volunteers were found to have blood alcohol levels of 104, 79, 88, 120 and 90 mg 100 ml^{-1}. A further set of six volunteers drank three pints of lager each, and were found to have blood alcohol levels of 68, 86, 71, 79, 91 and 66 mg 100 ml^{-1}. Use Tukey's quick test or the Mann–Whitney U-test to investigate the suggestion that drinking lager produces a lower blood alcohol level than drinking the same amount of beer. Use the Siegel–Tukey method to check whether the spreads of these two sets of results differ significantly.

6. A university chemical laboratory contains seven atomic-absorption spectrometers (A–G). Surveys of the opinions of the research students and of the academic staff show that the students' order of preference for the instruments is B, G, A, D, C, F, E, and that the staff members' order of preference is G, D, B, E, A, C, F. Are the opinions of the students and the staff correlated?

7. Use Theil's incomplete method to calculate the regression line for the data of Exercise 1 in Chapter 5.

8. Taking the radial immunodiffusion data from Exercise 3 above, use the Kolmogorov method to test the hypothesis that serum immunoglobulin G levels are normally distributed with a mean of 1.0 g 100 ml^{-1} and a standard deviation of 0.2 g 100 ml^{-1}. Will any other normal distribution fit the data better?

9. The nickel levels in three samples of crude oil were determined (six replicates in each case) by atomic-absorption spectrometry, with the following results:

Sample	*Measurements (Ni, ppm)*					
1	14.2	16.8	15.9	19.1	15.5	16.0
2	14.5	20.0	17.7	18.0	15.4	16.1
3	18.3	20.1	16.9	17.7	17.9	19.3

Use the Kruskal–Wallis method to decide whether the nickel levels in the three oils differ significantly.

CHAPTER SEVEN

Experimental design and optimization

7.1 Introduction

A recurring theme in this book has been that statistical methods are invaluable not only in the analysis of experimental data, but also in designing and optimizing experiments. So many experiments fail in their purpose because they are not properly thought out and designed in the first place, and in that case even the best data analysis procedures cannot compensate for such a fatal lack of foresight and planning. This chapter introduces the basic concepts of experimental design and optimization, and summarizes the methods that should be carefully considered and used before any new experimental procedure is started.

In Chapter 3 we introduced the idea of a **factor**, i.e. any aspect of the experimental conditions which affects the result obtained from an experiment. Section 3.9 gave the example of the dependence of a fluorescence signal on the conditions under which a solution was stored. The factor of interest was these storage conditions; it was called a **controlled factor** because it could be altered at will by the experimenter. In another example in Section 4.3, in which salt from different parts of a barrel was tested for purity, the factor of interest, i.e. the part of the barrel from which the salt was taken, was chosen at random, so that factor was called an **uncontrolled factor**. In both these examples the factors were **qualitative** since their possible 'values' could not be arranged in numerical order. A factor for which the possible values can be arranged in numerical order, e.g. temperature, is a **quantitative** one. The different values that a factor takes are known as different **levels**.

As presented in Chapters 3 and 4, these experiments were intended as an introduction to the calculations involved in the analysis of variance (ANOVA). No mention was made of other conditions that might have introduced further factors affecting the results, so experimental designs could not be considered. But in reality the fluorescence experiment might be affected by additional factors such as the ambient temperature, the use of the same or a different fluorescence spectrometer for each measurement, and the dates, times and experimenters used in making the measurements. Any of these factors might have influenced the results to give the observed behaviour, thus invalidating the conclusions concerning the effect of the storage conditions. Clearly, if the correct conclusions are to be drawn from an experiment, the various factors affecting the result must be identified in advance and, if possible, controlled.

The term **experimental design** is usually used to describe the stages of:

1. identifying the factors which may affect the result of an experiment;
2. designing the experiment so that the effects of uncontrolled factors are minimized;
3. using statistical analysis to separate and evaluate the effects of the various factors involved.

Since many factors will affect experimental results, quite complex experimental designs may be necessary. The choice of the best practical levels of these factors, i.e. the optimization of the experimental conditions, will also require detailed study. These methods, along with other multivariate methods covered in the next chapter, are amongst those given the general term **chemometrics**.

7.2 Randomization and blocking

One of the assumptions of one-way (and other) ANOVA calculations is that the uncontrolled variation is truly random. However, in measurements made over a period of time, variation in an uncontrolled factor such as pressure, temperature, deterioration of apparatus, etc., may produce a trend in the results. As a result the errors due to uncontrolled variation are no longer random since the errors in successive measurements are *correlated*. This can lead to a systematic error in the results. Fortunately this problem is simply overcome by using the technique of **randomization**. Suppose we wish to compare the effect of a single factor, the concentration of perchloric acid in aqueous solution, at three different levels or treatments (0.1 M, 0.5 M, and 1.0 M) on the fluorescence intensity of quinine (which is widely used as a primary standard in fluorescence spectrometry). Let us suppose that four replicate intensity measurements are made for each treatment, i.e. in each perchloric acid solution. Instead of making the four measurements in 0.1 M acid, followed by the four in 0.5 M acid, then the four in 1 M acid, we make the 12 measurements in a random order, decided by using a table of random numbers. Each treatment is assigned a number for each replication as follows:

0.1 M				0.5 M				1 M			
01	02	03	04	05	06	07	08	09	10	11	12

(Note that each number has the same number of digits.) We then enter a random number table (see Table A.8) at an arbitrary point and read off pairs of digits, discarding the pairs 00, 13–99 and also discarding repeats. Suppose this gives the sequence 02, 10, 04, 03, 11, 01, 12, 06, 08, 07, 09, 05. Then, using the numbers assigned above, the measurements would be made at the different acid levels in the following order: 0.1 M, 1 M, 0.1 M, 0.1 M, 1 M, 0.1 M, 1 M, 0.5 M, 0.5 M, 0.5 M, 1 M, 0.5 M. This random order of measurement ensures that the errors at each acid level due to uncontrolled factors are random.

One disadvantage of complete randomization is that it fails to take advantage of any natural sub-divisions in the experimental material. Suppose, for example, that all the 12 measurements in this example could not be made on the same day but were divided between four consecutive days. Using the same order as before would give:

Day 1	0.1 M,	1 M,	0.1 M
Day 2	0.1 M,	1 M,	0.1 M
Day 3	1 M,	0.5 M,	0.5 M
Day 4	0.5 M,	1 M,	0.5 M

With this design all the measurements using 0.1 M perchloric acid as the quinine solvent occur (by chance) on the first two days, whereas those using 0.5 M perchloric acid happen to be made on the last two days. If it seemed that there was a difference between the effects of these two acid levels it would not be possible to tell whether this difference was genuine or was caused by the effect of using the two treatments on different pairs of days. A better design is one in which each treatment is used once on each day, with the order of the treatments randomized on each day. For example:

Day 1	0.1 M,	1 M,	0.5 M
Day 2	0.1 M,	0.5 M,	1 M
Day 3	1 M,	0.5 M,	0.1 M
Day 4	1 M,	0.1 M,	0.5 M

A group of results which contains one measurement for each treatment (here, the measurements on each day) is known as a **block**, so this design is called a **randomized block** design. Further designs that do not use randomization at all are considered in Section 7.4.

7.3 Two-way ANOVA

When two factors may affect the results of an experiment, two-way Analysis of Variance must be used to study their effects. Table 7.1 shows the general

Table 7.1 General form of table of two-way ANOVA

	Treatment						*Row total*
	1	*2*	...	*j*	...	*c*	
Block 1	x_{11}	x_{12}	...	x_{1j}	...	x_{1c}	$T_{1.}$
Block 2	x_{21}	x_{22}	...	x_{2j}	...	x_{2c}	$T_{2.}$
.	.	.	...	.	...	.	.
Block i	x_{i1}	x_{i2}	...	x_{ij}	...	x_{ic}	$T_{i.}$
.	.	.	...	.	...	.	.
Block r	x_{r1}	x_{r2}		x_{rj}		x_{rc}	$T_{r.}$
Column total	$T_{\cdot 1}$	$T_{\cdot 2}$		$T_{\cdot j}$		$T_{.c}$	$T =$ grand total

form of a layout for this method. Each of the N measurements, x_{ij}, is classified under the terms **treatment levels** and **blocks**: the latter term was introduced in the previous section. (These terms are derived from the original use of ANOVA by R. A. Fisher in agricultural experiments, but are still generally adopted.) Using the conventional symbols there are c treatment levels and r blocks, so $N = cr$. The row totals ($T_{1.}$, $T_{2.}$, etc.) and the column totals ($T_{.1}$, $T_{.2}$, etc.), and the grand total, T, are also given as they are used in the calculations. (The dots in the column and row totals remind us that in each case only one of the two factors is being studied.) The formulae for calculating the variation from the three different sources, viz. between-treatment, between-block, and experimental error, are given in Table 7.2. Their derivation will not be given in detail here: the principles are similar to those for one-way ANOVA (Section 3.9) and the texts listed in the Bibliography provide further details.

As in one-way ANOVA, the calculations are simplified by the repeated appearance of the term T^2/N, and by the fact that the residual (random experimental) error is obtained by subtraction. Note that an estimate of this experimental error can be obtained, even though only one measurement is made at each combination of treatment level and block (e.g. each chelating agent is tested only once on each day in the example below).

Table 7.2. Formulae for two-way ANOVA

Source of variation	*Sum of squares*	*Degrees of freedom*
Between-treatment	$\sum T_{.j}^2/r - T^2/N$	$c - 1$
Between-block	$\sum T_{i.}^2/c - T^2/N$	$r - 1$
Residual	by subtraction	by subtraction
Total	$\sum\sum x_{ij}^2 - T^2/N$	$N - 1$

EXAMPLE 7.3.1

In an experiment to compare the percentage efficiency of different chelating agents in extracting a metal ion from aqueous solution the following results were obtained:

	Chelating agent			
Day	*A*	*B*	*C*	*D*
1	84	80	83	79
2	79	77	80	79
3	83	78	80	78

On each day a fresh solution of the metal ion (with a specified concentration) was prepared and the extraction performed with each of the chelating agents taken in a random order.

In this experiment the use of different chelating agents is a controlled factor since the chelating agents are chosen by the experimenter. The day on which the experiment is performed introduces uncontrolled variation, caused both by changes in laboratory temperature, pressure, etc., and slight differences in the concentration of the metal ion solution, i.e. the day is a random factor. In previous chapters it was shown that ANOVA can be used either to test for a significant effect due to a controlled factor, or to estimate the variance of an uncontrolled factor. In this case, where both types of factor occur, two-way ANOVA can be used in both ways: (i) to test whether the different chelating agents have significantly different efficiencies; and (ii) to test whether the day-to-day variation is significantly greater than the variation due to the random error of measurement and, if it is, to estimate the variance of this day-to-day variation. As in one-way ANOVA, the calculations can be simplified by subtracting an arbitrary number from each measurement. The table below shows the measurements with 80 subtracted from each.

	Treatments					
Blocks	*A*	*B*	*C*	*D*	*Row totals, $T_{i.}$*	$T_{i.}^2$
Day 1	4	0	3	−1	6	36
Day 2	−1	−3	0	−1	−5	25
Day 3	3	−2	0	−2	−1	1 $\sum T_{i.}^2 = 62$
Column totals, $T_{.j}$	6	−5	3	−4	0 = Grand total, T	
$\sum T_{.j}^2 = 86$	36	25	9	16		

We also have $r = 3$, $c = 4$, $N = 12$, and $\sum\sum x_{ij}^2 = 54$.

The calculation of the ANOVA table gives the following results:

Source of variation	*Sum of squares*	*d.f.*	*Mean square*
Between-treatment	$86/3 - 0^2/12 = 28.6667$	3	$28.6667/3 = 9.5556$
Between-block	$62/4 - 0^2/12 = 15.5$	2	$15.5/2 = 7.75$
Residual	by subtraction = 9.8333	6	$9.8333/6 = 1.6389$
Total	$54 - 0^2/12 = 54.0$	11	

It is important to note that, since the residual mean square is obtained by subtraction, many significant figures should initially be used in the table to avoid significant errors in cases where this calculated difference is small.

It is instructive to verify that this calculation does indeed separate the between-treatment and between-block effects. For example, if all the values in one block are increased by a fixed amount and the sums of squares re-calculated, it is found that, while the between-block and total sums of squares are changed, the between-treatment and residual sums of squares are not.

If there is no difference between the efficiencies of the chelators, and no day-to-day variation, then all three mean squares should give an estimate of σ_0^2, the variance of the random variation due to experimental error (see Section 3.9). As in one-way ANOVA, the *F*-test is used to see whether the variance estimates differ significantly. Comparing the between-treatment mean square with the residual mean square gives:

$$F = 9.5556/1.6389 = 5.83$$

From Table A.3 the critical value of $F_{3,6}$ (1-tailed, $P = 0.05$) is 4.76, so we find that there is a difference between the two variances, i.e. between the efficiency of the chelating agents, at the 5% level. Comparing the between-block (i.e. between-day) and residual mean squares gives:

$$F = 7.75/1.6389 = 4.73$$

Here the critical value is 5.14, so there is no significant difference between days. Nevertheless the between-block mean square is considerably larger than the residual mean square, and had the experiment been 'unblocked', so that these two effects were combined in the estimate of experimental error, the experiment would probably have been unable to detect whether different *treatments* gave significantly different results. If the difference between days *had* been significant it would indicate that other factors such as temperature, pressure, the preparation of the solution, etc., were having an effect. It can be shown that the between-block mean square gives an estimate of $\sigma_0^2 + c\sigma_b^2$,where σ_b^2 is the variance of the random day-to-day variation. Since the residual mean square gives an estimate of σ_0^2, an estimate of σ_b^2 can be obtained.

This example illustrates clearly the benefits of considering carefully the design of an experiment before it is performed. Given a blocked and an unblocked experiment with the same number of measurements in each, the former is more sensitive and yields more information. The sensitivity of the experiment depends on the size of the random variation: the smaller this is, the smaller the difference between the treatments that can be detected. In an unblocked experiment the random variation would be larger since it would include a contribution from the day-to-day variation, so the sensitivity would be reduced.

The two-way ANOVA calculation performed above is based on the assumption that the effects of the chelators and the days, if any, are *additive*, not *interactive*. This point is discussed further in Section 7.5.

7.4 Latin squares and other designs

In some experimental designs it is possible to take into account an extra factor without a large increase in the number of experiments performed. A simple example is provided by the study of the chelating agents in the previous section, where an uncontrolled factor not taken into account was the time of day at which the measurements were made. Systematic variation during the day due to deterioration of the solutions or an increase in laboratory temperature could have produced a trend in the results.

In such cases, when there is an *equal number* of blocks and treatments (this was not the case in the previous example) it is possible to use an experimental design which allows the separation of such an additional factor. Suppose that the treatments are simply labelled A, B, and C, then a possible design would be:

Day 1	A	B	C
Day 2	C	A	B
Day 3	B	C	A

This block design, in which each treatment appears once in each row and once in each column, is known as a **Latin square**. It allows the separation of the variation into the between-treatment, between-block, between-time-of-day and random experimental error components. More complex designs are possible which remove the constraint of equal numbers of blocks and treatments. If there are more than three blocks and treatments a number of Latin square designs are obviously possible (one can be chosen at random). Experimental designs of the types discussed so far are said to be **cross-classified** designs, as they provide for measurements for every possible combination of the factors. But in other cases (for example when samples are sent to different laboratories, and are analysed by two or more different experimenters in each laboratory) the designs are said to be **nested** or **hierarchical**, because the experimenters do not make measurements in laboratories other than their own. Mixtures between nested and cross-classified designs are also possible.

7.5 Interactions

In the example in Section 7.3 we saw that the two-way ANOVA calculations used assumed that the effects of the two factors (chelating agents and days) were additive. This means that if, for example, we had had only two chelating agents, A and B, and studied them both on each of two days, the results might have been something like:

	Chelating agents	
	A	*B*
Day 1	80	82
Day 2	77	79

That is, using chelating agent B instead of A produces an increase of 2% in extraction efficiency on both days; and the extraction efficiency on day 2 is lower than that on day 1 by 3%, whichever chelating agent is used. In a simple table of the kind shown, this means that when three of the measurements are known, the fourth can easily be deduced. Suppose, however, that the extraction efficiency on day 2 for chelating agent B had been 83% instead of 79%. Then we would conclude that the difference between the two agents depended on the day of the measurements, or that the difference between the results on the two days depended on which agent was in use. That is, there would be an **interaction** between the two factors affecting the results. Such interactions are in practice extremely important: a recent estimate suggests that at least two-thirds of the processes in the chemical industry are affected by interacting, as opposed to additive, factors.

Unfortunately the detection of interactions is not quite so simple as the above example implies, as the situation is confused by the presence of random errors. If a two-way ANOVA calculation is applied to the very simple table above, the residual sum of squares will be found to be zero, but if any of the four values is altered this is no longer so. With this design of experiment we cannot tell whether a non-zero residual sum of squares is due to random errors, to an interaction between the factors, or to both effects. To resolve this problem the measurements in each cell must be replicated. The manner in which this is done is important: the measurements must be repeated in such a way that all the sources of random error are present in every case. Thus in our example if different glassware or other equipment items have been used in experiments on the different chelating agents, then the replicate measurements applied to each chelating agent on each day must also use different apparatus. If the same equipment is used for these replicates, clearly the random error in the measurements will be underestimated. If the replicates are performed properly the method by which the interaction sum of squares and the random error can be separated is illustrated by the following new example.

EXAMPLE 7.5.1

In an experiment to investigate the validity of a solution as a liquid absorbance standard, the value of the molar absorptivity, ε, of solutions of three different concentrations was calculated at four different wavelengths. Two replicate absorbance measurements were made for each combination of concentration and wavelength. The order in which the measurements were made was randomized. The results are shown in Table 7.3: for simplicity of calculation the calculated ε values have been divided by 100.

Table 7.4 shows the result of the Minitab calculation for these results. (NB. In using this program for two-way ANOVA calculations with interaction, it is essential to avoid the option for an additive model: the latter excludes the desired interaction effect. Excel also provides facilities for including interaction effects in two-way ANOVA.) Here we explain in more detail how this ANOVA table is obtained.

Table 7.3 Molar absorptivity values for a possible absorbance standard

	Wavelength, nm			
Concentration, g l^{-1}	*240*	*270*	*300*	*350*
0.02	94, 96	106, 108	48, 51	78, 81
0.04	93, 93	106, 105	47, 48	78, 78
0.06	93, 94	106, 107	49, 50	78, 79

Table 7.4 Minitab output for Example 7.5.1

```
Two-way analysis of variance

Analysis of Variance for Response
Source          DF         SS        MS
Conc.            2      12.33      6.17
Wavelength       3   11059.50   3686.50
Interaction      6       2.00      0.33
Error           12      16.00      1.33
Total           23   11089.83
```

The first stage of the calculation is to find the cell totals. This is done in Table 7.5, which also includes other quantities needed in the calculation. As before, $T_{i.}$ denotes the total of the *i*th row, $T_{.j}$ the total of the *j*th column and *T* the grand total.

Table 7.5 Cell totals for two-way ANOVA calculation

	240 nm	*270 nm*	*300 nm*	*350 nm*	$T_{i.}$	$T_{i.}^2$
0.02 g l^{-1}	190	214	99	159	662	438244
0.06 g l^{-1}	186	211	95	156	648	419904
0.10 g l^{-1}	187	213	99	157	656	430336
$T_{.j}$	563	638	293	472	$T = 1966$	
$T_{.j}^2$	316969	407044	85849	222784		

$$\sum_j T_{.j}^2 = 1032646 \qquad \sum_i T_{i.}^2 = 1288484$$

As before, the between-row, between-column and total sums of squares are calculated. Each calculation requires the term T^2/nrc (where *n* is the number of replicate measurements in each cell, in this case 2, *r* is the number of rows and c is the number of columns). This term is sometimes called the **correction term,**

C. Here we have:

$$C = T^2/nrc = 1966^2/(2 \times 3 \times 4) = 161048.17$$

The sums of squares are now calculated:

$$\text{Between-row sum of squares} = \sum_i T_{i.}^2/nc - C$$
$$= 1288484/(2 \times 4) - 161048.17$$
$$= 12.33$$

with $r - 1 = 2$ degrees of freedom.

$$\text{Between-column sum of squares} = \sum_j T_{.j}^2/nr - C$$
$$= 1032646/(2 \times 3) - 161048.17$$
$$= 11059.50$$

with $c - 1 = 3$ degrees of freedom.

$$\text{Total sum of squares} = \sum x_{ijk}^2 - C$$

where x_{ijk} is the *k*th replicate in the *i*th row and *j*th column, i.e. $\sum x_{ijk}^2$ is the sum of the squares of the individual measurements in Table 7.3.

$$\text{Total sum of squares} = 172138 - 161048.17$$
$$= 11089.83$$

with $nrc - 1 = 23$ degrees of freedom.

The variation due to random error (usually called the **residual variation**) is estimated from the within-cell variation, i.e., the variation between replicates.

The residual sum of squares $= \sum x_{ijk}^2 - \sum T_{ij}^2/n$, where T_{ij} is the total for the cell in the *i*th row and *j*th column, i.e. the sum of the replicate measurements in the *i*th row and *j*th column.

$$\text{Residual sum of squares} = \sum x_{ijk}^2 - \sum T_{ij}^2/n$$
$$= 172138 - (344244/2)$$
$$= 16$$

with $(n - 1)rc = 12$ degrees of freedom.

The interaction sum of squares and number of degrees of freedom can now be found by subtraction. Each source of variation is compared with the residual mean square to test whether it is significant.

1. **Interaction.** This is obviously not significant since the interaction mean square is less than the residual mean square.
2. **Between-column** (i.e. between-wavelength). This is highly significant since we have:

$$F = 3686.502/1.3333 = 2765$$

The critical value of $F_{3,12}$ is 3.49 ($P = 0.05$). In this case a significant result would be expected since absorbance is wavelength-dependent.

3. **Between-row** (i.e. between-concentration). We have:

$$F = 6.17/1.3333 = 4.63$$

The critical value of $F_{2,12}$ is 3.885 ($P = 0.05$), indicating that the between-row variation is too great to be accounted for by random variation. So the solution is not suitable as an absorbance standard. Figure 7.1 shows the molar absorptivity plotted against wavelength, with the values for the same concentration joined by straight lines. This illustrates the results of the analysis above in the following ways:

1. The lines are parallel, indicating no interaction.
2. The lines are not quite horizontal, indicating that the molar absorptivity varies with concentration.
3. The lines are at different heights on the graph, indicating that the molar absorptivity is wavelength-dependent.

The formulae used in the calculation above are summarized in Table 7.6.

In this experiment both factors, i.e. the wavelength and the concentration of the solution, are controlled factors. In analytical chemistry an important application of ANOVA is the investigation of two or more controlled factors and their interactions in optimization experiments. This is discussed in Section 7.7.

As discussed in Section 4.11, another important application of ANOVA is in collaborative investigations of precision and accuracy between laboratories. In full scale collaborative trials several different types of sample are sent to a number of laboratories with each laboratory performing a number of replicate analyses on each sample. Mathematical analysis of the results would yield the following sums of squares: between-laboratory, between-samples, laboratory-sample interaction, and residual. The purpose of such an experiment would be to test first whether there is any interaction between laboratory and sample, i.e. whether some laboratories showed unexpectedly high or low results for some samples. This is done by comparing the interaction and residual

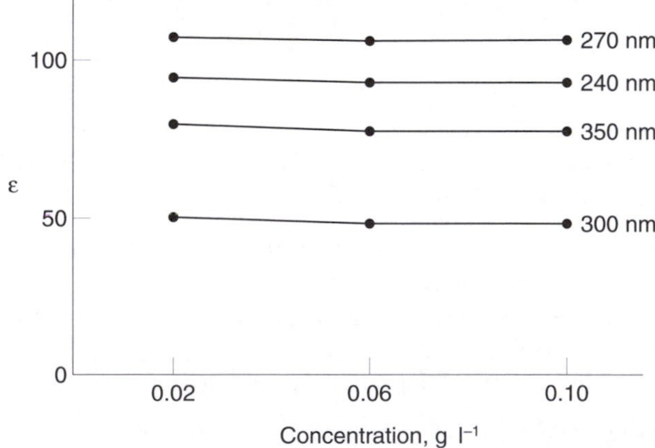

Figure 7.1 Relationships in the two-way ANOVA example (Example 7.5.1).

Table 7.6. Formulae for two-way ANOVA with interaction

Source of variation	*Sum of squares*	*Degrees of freedom*
Between-row	$\sum_i T_{i.}^2/nc - C$	$r-1$
Between-column	$\sum_j T_{.j}^2/nr - C$	$c-1$
Interaction	by subtraction	by subtraction
Residual	$\sum x_{ijk}^2 - \sum T_{ij}^2/n$	$rc(n-1)$
Total	$\sum x_{ijk}^2 - C$	$rcn-1$

sums of squares. If there is no interaction, then we could test whether the laboratories obtained significantly different results, i.e. if there is any systematic difference between laboratories. If there is, then the inter-laboratory variance can be estimated. However, if there *is* a significant interaction, the testing for a significant difference between laboratories has little relevance.

For two-way ANOVA to be valid the following conditions must be fulfilled (see also Section 3.10):

1. The random error is the same for all combinations of the levels of the factors.
2. The random errors are approximately normally distributed.

7.6 Factorial versus one-at-a-time design

An experiment such as the one in the previous example, where the response variable (i.e. the molar absorptivity) is measured for all possible combinations of the chosen factor levels, is known as a **complete factorial design**. The reader may have noticed that this design of the experiment is the antithesis of the classical approach in which the response is investigated for each factor in turn while all the other factors are held at a constant level. There are two reasons for preferring a factorial design to a classical design in experiments that test whether the response depends on factor level:

1. The factorial experiment detects and estimates any interaction, which the one-at-a-time experiment cannot.
2. If the effects of the factors are additive, then the factorial design needs fewer measurements than the classical approach in order to give the same precision. This can be seen by turning again to the molar absorptivity experiment. There, all 24 measurements were used to estimate the effect of varying the wavelength and the *same* 24 were used to estimate the effect of varying concentration. In a one-at-a-time experiment, first the concentration would have been fixed and, to obtain the same precision for the effect of varying the wavelength, six measurements would have been needed at each wavelength, i.e. 24 in all. Then the wavelength would have been fixed and another 24 measurements made at different concentrations, making a total of 48 altogether. In general, for k factors, a classical approach involves k times as many measurements as a factorial one with the same precision.

7.7 Factorial design and optimization

In many analytical techniques the response of the measurement system depends on a variety of experimental factors under the control of the operator. For example, enzyme assays involve the direct or indirect measurement of reaction rates. In a given experiment the reaction rate will depend on factors such as the temperature, the pH, ionic strength, and chemical composition of the buffer solution, the enzyme concentration, and so on. For a particular application it will be important to set the levels of these factors to ensure that (for example) the reaction rate is as high as possible. The process of finding these optimum factor levels is known as **optimization.** Several methods of optimization are discussed in detail in subsequent sections. But before an optimization process can begin we must determine which factors, and which interactions between them, are important in affecting the response: it is also valuable to know which factors have little or no effect, so that time and resources are not wasted on unnecessary experiments.

Such studies use a factorial experiment with each factor at two levels usually known as 'low' and 'high'. In the case of a quantitative variable the terms 'low' and 'high' have their usual meaning. The exact choice of levels is determined principally by the experience and knowledge of the experimenter and the physical constraints of the system, e.g. in aqueous solutions only temperatures in the range 0–100°C are practicable. Some problems affecting the choice of levels are discussed below. For a qualitative variable 'high' and 'low' refer to a pair of different conditions, e.g. the presence or absence of a catalyst, use of mechanical or magnetic stirring, a sample in powdered or granular form, etc. Since we have already considered two-factor experiments in some detail we will turn to one with three factors: A, B and C. This means that there are $2 \times 2 \times 2 = 8$ possible combinations of factor levels, as shown in the table below. A plus sign denotes that the factor is at the high level and a minus sign that it is at the low level. The first column gives a notation often used to describe the combinations, where the presence of the appropriate lower case letter indicates that the factor is at the high level and its absence that the factor is at the low level. The number 1 is used to indicate that all factors are at the low level.

Combination	*A*	*B*	*C*	*Response*
1	−	−	−	y_1
a	+	−	−	y_2
b	−	+	−	y_3
c	−	−	+	y_4
ab	−	+	+	y_5
ac	+	−	+	y_6
bc	+	+	−	y_7
abc	+	+	+	y_8

The method by which the effects of the factors and their interactions are estimated is illustrated by the following example.

EXAMPLE 7.7.1

In a high-performance liquid chromatography experiment, the dependence of the retention parameter, k', on three factors was investigated. The factors were pH (factor P), the concentration of a counter-ion (factor T) and the concentration of the organic solvent in the mobile phase (factor C). Two levels were used for each factor and two replicate measurements made for each combination. The measurements were randomized. The table below gives the average value for each pair of replicates.

Combination of factor levels	k'
1	4.7
p	9.9
t	7.0
c	2.7
pt	15.0
pc	5.3
tc	3.2
ptc	6.0

Effect of individual factors

The effect of changing the level of P can be found from the average difference in response when P changes from high to low level with the levels of C and T fixed. There are four pairs of responses that give an estimate of the effect of the level of P as shown in the table below.

Level of C	*Level of T*	*Level of P*		*Difference*
		+	−	
−	−	9.9	4.7	5.2
+	−	5.3	2.7	2.6
−	+	15.0	7.0	8.0
+	+	6.0	3.2	2.8
			Total =	18.6

Average effect of altering the level of P = 18.6/4 = 4.65

The average effects of altering the levels of T and C can be found similarly to be:

Average effect of altering the level of C = −4.85
Average effect of altering the level of T = 2.15

Interaction between two factors

Consider now the two factors P and T. If there is no interaction between them then the change in response between the two levels of P should be independent of the level of T. The first two figures in the last column of the table above give the change in response when P changes from high to low level with T at low level.

Their average is $(5.2 + 2.6)/2 = 3.9$. The last two figures in the same column give the effect of changing P when T is at high level. Their average is $(8.0 + 2.8)/2 = 5.4$. If there is no interaction and no random error (see Section 7.5) these estimates of the effect of changing the level of P should be equal. The convention is to take half their difference as a measure of the interaction:

$$\text{Effect of PT interaction} = (5.4 - 3.9)/2 = 0.75$$

It is important to realize that this quantity estimates the degree to which the effects of P and T are not additive. It could equally well have been calculated by considering how far the change in response for the two levels of T is independent of the level of P.

The other interactions are calculated in a similar fashion:

$$\text{Effect of CP interaction} = -1.95$$
$$\text{Effect of CT interaction} = -1.55$$

Interaction between three factors

The PT interaction calculated above can be split into two parts according to the level of C. With C at low level the estimate of interaction would be $(8.0 - 5.2)/2 = 1.4$ and with C at high level it would be $(2.8 - 2.6)/2 = 0.1$. If there is no interaction between all three factors and no random error, these estimates of the PT interaction should be equal. The three-factor interaction is estimated by half their difference $[= (0.1 - 1.4)/2 = -0.65]$. The three-factor interaction measures the extent to which the effect of the PT interaction and the effect of C are not additive: it could equally well be calculated by considering the difference between the PC estimates of interaction for low and high levels of T or the difference between the TC estimates of interaction for low and high levels of P.

These results are summarized in the table below.

	Effect
Single factor (main effect)	
P	4.65
T	2.15
C	−4.85
Two-factor interactions	
TP	0.75
CT	−1.55
CP	−1.95
Three-factor interactions	
PTC	−0.65

These calculations have been presented in some detail in order to make the principles clear. An algorithm due to Yates (see Bibliography) simplifies the calculation.

In order to test which effects, if any, are significant, ANOVA may be used (provided that there is homogeneity of variance). It can be shown that in a two-level experiment, like this one, the required sums of squares can be calculated from the estimated effects by using

$$\text{Sum of squares} = N \times (\text{estimated effect})^2/4$$

where N is the total number of measurements, including replicates. In this case N is 16 since two replicate measurements were made for each combination of factor levels. The calculated sums of squares are given below.

Factor(s)	*Sum of squares*
P	86.49
T	18.49
C	94.09
PT	2.25
TC	9.61
PC	15.21
PCT	1.69

It can be shown that each sum of squares has one degree of freedom and since the mean square is given by

$$\text{mean square} = \text{sum of squares/number of degrees of freedom}$$

each mean square is simply the corresponding sum of squares. To test for the significance of an effect, the mean square is compared with the error (residual) mean square. This is calculated from the individual measurements by the method described in the molar absorptivity example in Section 7.5. In the present experiment the calculated residual mean square was 0.012 with eight degrees of freedom. Testing for significance, starting with the highest order interaction, we have for the PTC interaction:

$$F = 1.69/0.012 = 141$$

which is obviously significant. If there is interaction between all three factors there is no point in testing whether the factors taken in pairs or singly are significant, since all factors will have to be considered in any optimization process. A single factor should only be tested for significance if it does not interact with other factors.

One problem with a complete factorial experiment such as this is that the number of experiments required rises rapidly with the number of factors: for k factors at two levels with two replicates for each combination of levels, 2^{k+1} experiments are necessary, e.g. for five factors, 64 experiments. When there are more than three factors some economy is possible by assuming that three-way and higher order interactions are negligible. The sums of squares corresponding to these interactions can then be combined to give an estimate of the residual sum of squares, and replicate measurements are no longer necessary. The rationale for this approach is that higher order effects are usually much smaller than main effects and two-factor interaction effects. If higher order interactions can be assumed negligible, a suitable fraction of all possible combinations of factor levels is sufficient to provide an estimate of the main and two-factor interaction effects. As mentioned in Section 4.11, such an experimental design is called an **incomplete** or **fractional factorial design**.

Another problem in using a factorial design to determine which factors have a significant effect on the response is that, for factors which are continuous variables, the effect depends on the high and low levels used. If the high and low levels are too close together the effect of the corresponding factor may be found not significant despite the fact that over the whole possible range of factor levels the effect of this factor is not negligible. On the other hand, if the levels are too far apart they may fall on either side of a maximum and still give a difference in response that is not significant.

7.8 Optimization: basic principles and univariate methods

When the various factors and interactions affecting the results of an experiment have been identified, separate methods are needed to determine the combination of factor levels which will provide the optimum response. It is first necessary to define carefully what is meant by the 'optimum response' in a given analytical procedure. In some cases the aim will be to ensure that the measuring instrument gives a *maximum* response signal, e.g. the largest possible absorbance, current, emission intensity, etc. However in many other cases the optimum outcome of an experiment may be the maximum signal to noise or signal to background ratios, the best resolution (in separation methods), or even a minimum response (e.g. when an interfering signal is under study). In mathematical terms, finding maxima and minima are virtually identical processes, so the last example causes no additional problems. It may be stating the obvious to emphasize that the exact aim of an optimization experiment must be carefully defined in advance, but in practice many optimization processes have failed simply because the target was not sufficiently clearly laid down.

A good optimization method has two qualities. It produces a set of experimental conditions that provide the optimum response, or at least a response that is close to the optimum: and it does so with the smallest possible number of trial experimental steps. In practice the speed and convenience of the optimization procedure is extremely important, and it may be sufficient in some cases to use a method that gets reasonably close to the true optimum in a small number of steps.

In this context we must note that even the optimization of a single factor presents some interesting problems. Suppose we wish to find the optimum pH of an enzyme catalysed reaction within the pH range 2–12, the best pH being that at which the reaction rate is a maximum. Each reaction rate measurement will be an entirely separate experiment, taking significant time

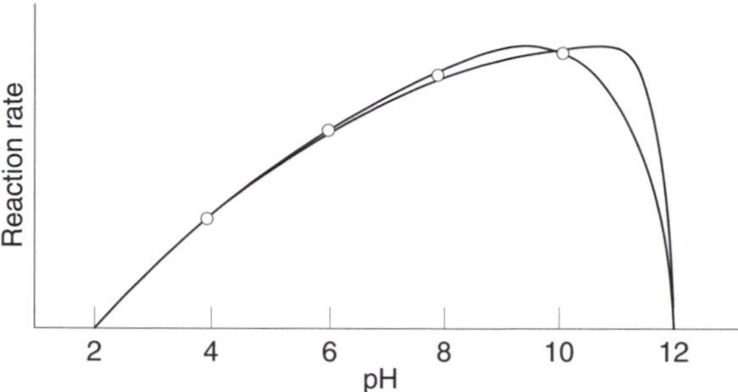

Figure 7.2 Optimization experiment with equally spaced factor levels.

and effort and with a different buffer solution in each case, so it is particularly important to get as much information as possible from the smallest possible number of experiments. Two approaches suggest themselves. One is to make a fixed number of measurements of the reaction rate, for example by dividing the pH interval of interest up into a number of equal regions. The second and more logical method is to make the measurements sequentially, so that the pH for each experiment depends on the results of the previous experiments.

Figure 7.2 shows the result of making four rate measurements at pH values of 4, 6, 8, and 10. In considering these four results we shall make an assumption that will underlie much of our discussion of optimization, i.e. that there is only one maximum within the range of the factor levels under study. (Inevitably, this is not always true, and we return to the point later.) The four points on the graph represent the results of the experiments: the highest reaction rate is obtained at pH 10, and the next highest at pH 8. But even with the assumption of a single maximum it is possible to draw two types of curve through the points, i.e. this maximum may occur between pHs 8 and 10, or between pHs 10 and 12. So the result of the four experiments is that, starting with the pH range between 2 and 12, we conclude that the optimum pH is actually between 8 and 12, i.e. we have narrowed the possible range for the optimum by a factor of 4/10. This is an example of the general result that, if n experiments are done using equal intervals of the factor level, the range for the optimum is narrowed by a factor of $2/(n+1)$ or 2/5 here. This is not a very impressive result! The weakness of the method is further shown by the fact that, if we wished to define the optimum pH within a range of 0.2 units, i.e. a 50-fold reduction of the original range of 10 units, 99 experiments would be needed, an obvious impossibility.

The principle of the superior step-wise approach is shown in Figure 7.3, which shows a possible relationship between reaction rate and pH. (This curve would of course not be known in advance to the experimenter.) In brief, the procedure is as follows. The first two experiments are carried out at pHs A and B, equidistant from the extremes of the pH range, 2 and 12. (The choice of pH values for these first experiments is discussed below.) The experiment at B will give the higher reaction rate so, since there is only

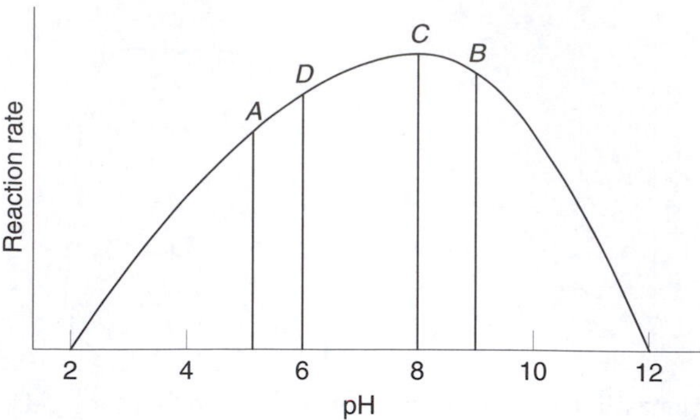

Figure 7.3 Step-wise approach to univariate searching.

one maximum in the curve, the portion of the curve between pH 2 and A can be rejected. The remainder of the pH range, between A and pH 12, certainly includes the maximum, and it already has one reading, B, within it. A new measurement, C, is then made at a pH such that the pH difference between C and A is the same as that between B and pH 12. The pH at C gives a higher reaction rate than B, so the interval between B and pH 12 can now be rejected, and a new measurement, D, made so that the A–D and C–B distances are equal. Further measurements use the same principle, so it only remains to establish how many steps are necessary, and where the starting points A and B should be.

In one approach the distances between the pairs of measurements and the extremes of the corresponding ranges are related to the **Fibonacci series**. This series of numbers, known since the thirteenth century, starts with 1 and 1 (these terms are called F_0 and F_1), with each subsequent term being the sum of the two previous ones. Thus F_2, F_3 etc. are 2, 3, 5, 8, 13, 21, 34, 55, 89... To use this series to optimize a single factor over a defined range we begin by deciding *either* on the degree of optimization required, which automatically determines the number of experiments necessary, *or* on the number of experiments we can perform, which automatically determines the degree of optimization obtained. Suppose that, as before, we require the optimum pH to be known within 0.2 units, a 50-fold reduction of the original pH interval of 10 units. We must then take the first Fibonacci number above 50: this is 55, F_9. The subscript tells us that nine experiments will be needed to achieve the desired result. The spacing of the first two points, A and B, within the range, is also given by the series. We use F_9 and the member of the series two below it, F_7, to form the fraction F_7/F_9, i.e. 21/55. Point A is then at pH $(2 + [10 \times 21/55])$, and point B is at pH $(12 - [10 \times 21/55])$, i.e. 5.8 and 8.2 respectively. (The number 10 appears in these expressions because the pH range of interest is 10 units in width.) Once these first points are established, the positions of C, D, etc. follow automatically by symmetry. It is striking that the Fibonacci search method achieves in just nine experiments a degree of optimization that requires 99 experiments using the 'equal intervals' method. It can be shown that the Fibonacci method is indeed the most efficient univariate search procedure for a given range when the degree of optimization is known or decided in advance.

In other optimization methods, it is not necessary to decide in advance either the number of experiments or the degree of optimization needed. The **Golden Ratio** or **Golden Section** method is an example of this approach. The Golden Ratio is $(1+\sqrt{5})/2 = 1.618$. This number has the property that $1.618 = 1 + (1/1.618)$. The first two points in the search are positioned by dividing the total range (e.g. 10 pH units) by 1.618, and placing the points the resulting distances from the two extremes of the range. In our example the result is that the points are placed $10/1.618 = 6.18$ pH units from pH 2 and pH 12, i.e. at pHs 5.82 and 8.18. If the latter point gives the higher response we can be sure that the optimum pH lies between 5.82 and 12, an interval already containing one result at pH 8.18. As in the Fibonacci method, the third experiment is then done at a pH placed symmetrically, i.e. at pH $(12 - [8.18 - 5.82]) = 9.64$. (Note that, once this third pH is established, $(12 - 8.18)/(12 - 9.64) = 1.618$, the Golden Ratio again.) If this third experiment gives a higher response than the experiment at pH 8.18, then it is apparent that the optimum is not found in the range pH 5.82–8.18, so the fourth experiment can be carried out at pH 10.54 (because $[12 - 10.54] = [9.64 - 8.18]$), and so on. Again we see that this method is more efficient than the 'equal intervals' approach, as after this fourth experiment the optimum pH will be placed in an interval of 2.36 units, i.e. either between pH 8.18 and pH 10.54, or between pH 9.64 and pH 12. This is significantly better, i.e. narrower, than the range of four pH units given by the equal intervals method with four experiments. The iteration can be continued until the analyst decides either that enough experiments have been performed, or that the optimization achieved is in practice good enough. The Fibonacci and Golden Ratio methods are similar and indeed mathematically related. Further details of univariate methods are given in the texts listed in the Bibliography.

It must be added that the success of this and other optimization procedures depends on the assumption that the random errors in the measurements (of reaction rates in our example) are significantly smaller than the rate of change of the response with the factor level (pH). This assumption is most likely to be invalid near to the optimum value of the response, where the slope of the response curve is near zero. This confirms that in many practical cases an optimization method which gets fairly close to the optimum in a few experimental steps will be most valuable: trying to refine the optimum by extra experiments might fail if the experimental errors give misleading results.

7.9 Optimization using the alternating variable search method

When the response of an analytical system depends on two factors which are continuous variables the relationship between the response and the levels of the two factors can be represented by a surface in three dimensions as

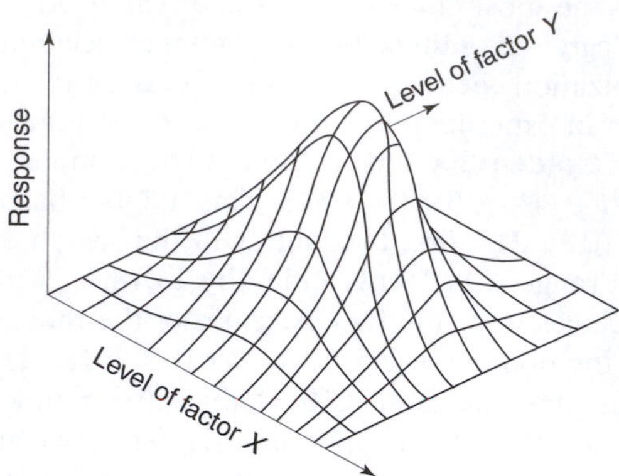

Figure 7.4 A response surface for two factors.

shown in Figure 7.4. This surface is known as the **response surface**, with the target optimum being the top of the 'mountain.' A more convenient representation is a **contour diagram** (Figure 7.5). Here the response on each contour is constant, and the target optimum is close to the centre of the contours. The form of the contour lines is, of course, unknown to the experimenter who wishes to determine the optimum levels, x_0 and y_0 for the factors X and Y respectively. A search method using a one-at-a-time approach would set the initial level of X to a fixed value at x_1, say, and vary the level of Y to give a maximum response at the point A, where the level of Y *is* y_1. Next, holding the level of Y at y_1 and varying the level of X would give a maximum at B. Obviously this is not the true maximum, as the position obtained depends on the initial value chosen for x_1. A better response can be obtained by

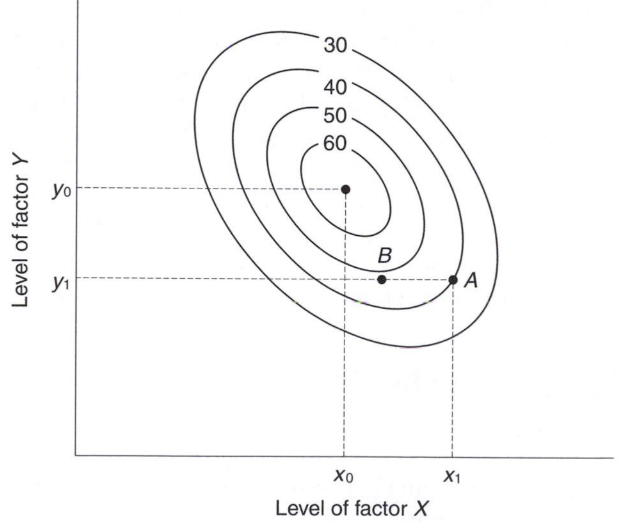

Figure 7.5 The contour diagram for a two-factor response surface.

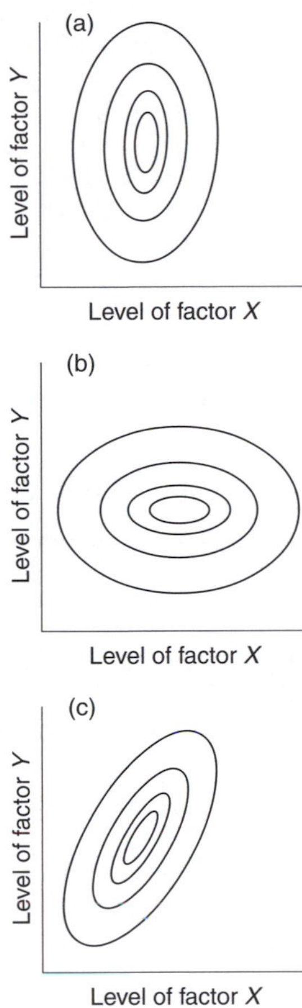

Figure 7.6 Simplified contour diagrams. (a) and (b) show no *X*–*Y* interaction; (c) shows significant *X*–*Y* interaction.

repeating the process, varying the levels of *X* and *Y* alternately. This method is known as the **alternating variable search** (AVS) or the **iterative univariate method**. When there is no interaction between the two factors this method is extremely efficient. In such a case the response surface has the form of Figure 7.6(a) or (b) and varying *X* and then *Y* just once will lead to the maximum response. If, however, there is interaction between the two variables then the response surface has the form of Figure 7.6(c) and *X* and *Y* must then be varied in turn a number of times. In some cases, even this will not lead to the true maximum: this is illustrated in Figure 7.7 where, although *C* is not the true maximum, the response falls on either side of it in both the *X* and the *Y* directions. A one-at-a-time method arriving at this point would therefore conclude wrongly that it represented the maximum response.

The AVS method has been used with success in some areas of analytical chemistry. But it is practicable only if the response can be monitored continuously as the factor level is altered easily, for example in spectrometry

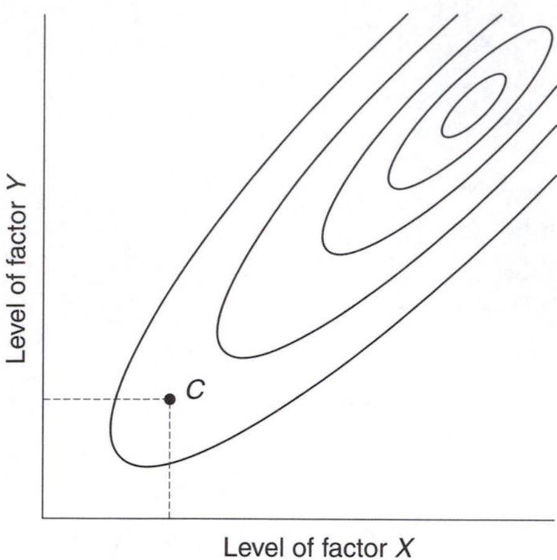

Figure 7.7 Contour diagram: a situation in which the one-at-a-time method fails to locate the maximum.

when the monochromator wavelength or slit width is readily changed. If such monitoring is not possible then a choice of step size has to be made for the change in each of the factors. A more sophisticated method would allow changes in these step sizes depending on the observed change in response, but in practice other optimization methods involving fewer separate experiments would be used.

There are obviously some problems of both principle and practice in applying a one-at-a-time approach to optimization, and a number of other methods have been developed in an attempt to overcome these problems and to minimize the number of experiments required. All these methods can be applied to any number of factors, but the response surface cannot easily be visualized for three or more factors: our remaining discussion of optimization methods will thus largely be confined to experiments involving two factors.

7.10 The method of steepest ascent

The process of optimization can be visualized in terms of a person standing on a mountain (Figure 7.4) in thick fog, with the task of finding the summit! In these circumstances an obvious approach is to walk in the direction in which the gradient is steepest. This is the basis of the **method of steepest ascent**. Figure 7.8 shows two possible contour maps. The direction of steepest ascent at any point is at right angles to the contour lines at that point, as indicated by the arrows. When the contour lines are circular this will be towards the summit but when the contour lines are elliptical it may not. The shape of the contour lines depends on the scales chosen for the axes: the best results are obtained from the method if the axes are scaled so that a unit change in either direction gives a roughly equal change in

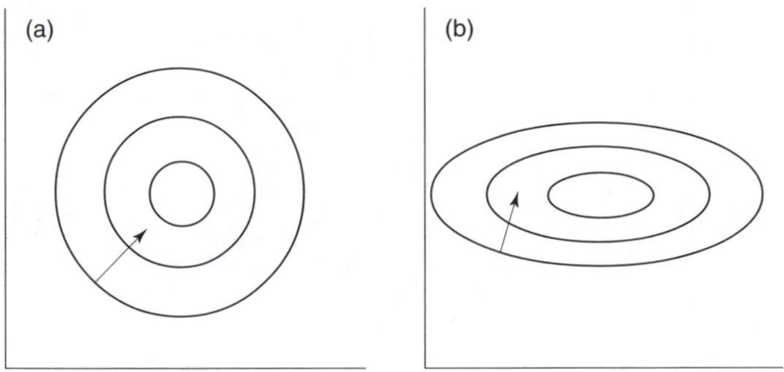

Figure 7.8 Contour diagrams: the arrow in each diagram indicates the path of steepest ascent. In (a) it goes close to the maximum but in (b) it does not.

response. The first step is to perform a factorial experiment with each factor at two levels. The levels are chosen so that the design forms a square as shown in Figure 7.9. Suppose, for example, that the experiment is an enzyme catalysed reaction in which the reaction rate, which in this case is the response, is to be maximized with respect to pH (X) and temperature (Y). The table below gives the results (reaction rate measured in arbitrary units) of the initial factorial experiment.

		pH (X)	
		6.8	*7.0*
Temperature, °C (Y)	20	30	35
	25	34	39

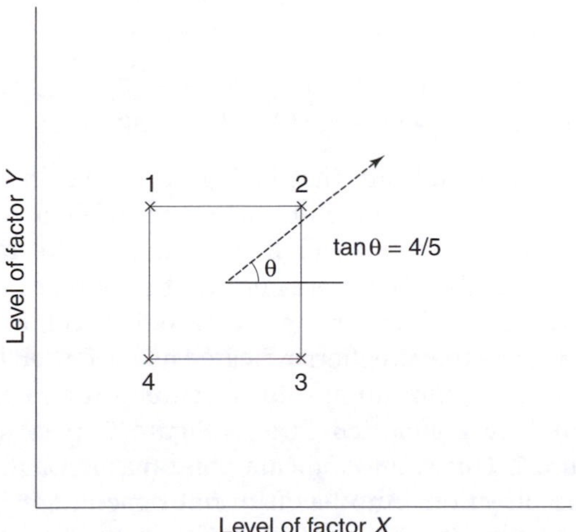

Figure 7.9 A 2 × 2 factorial design to determine the direction of steepest ascent, indicated by the broken line.

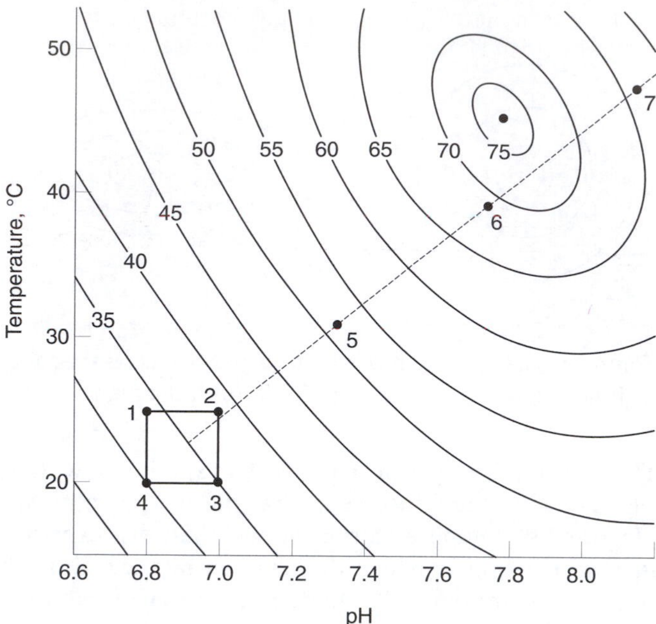

Figure 7.10 Contour diagram: the initial direction of steepest ascent is shown by the broken line. Further experiments are done at points 5, 6 and 7.

The effects of the two factors can be separated as described in Section 7.7. Rewriting the table above, in the notation of that section, gives:

Combination of levels	*Rate of reaction*
1	30
x	35
y	34
xy	39

Average effect of change in level of $X = [(35 - 30) + (39 - 34)]/2 = 5$
Average effect of change in level of $Y = [(34 - 30) + (39 - 35)]/2 = 4$

The effects of X and Y indicate that in Figure 7.9 we should seek for the maximum response to the right and above the original region. Since the change in the X direction is greater than in the Y direction the distance moved in the former should be greater. To be more exact, the distance moved in the X direction should be in the ratio 5 : 4 to the direction moved in the Y direction, i.e. in the direction indicated by the dotted line in Figure 7.9.

The next step in the optimization is to carry out further experiments in the direction indicated by the dotted line in Figure 7.10, at (say) the points numbered 5, 6 and 7. This would indicate point 6 as a rough position for the maximum in this direction. Another factorial experiment is carried out in this region to determine the new direction of steepest ascent.

This method gives satisfactory progress towards the maximum provided that, over the region of the factorial design, the contours are approximately straight. This is equivalent to the response surface being a plane which can be described mathematically by a linear combination of terms in x and y. Nearer the summit terms in xy, x^2 and y^2 are also needed to describe the surface. The xy term represents the interaction between X and Y and can be estimated by using replication as described in Section 7.7. The squared terms, which represent the curvature of the surface, can be estimated by comparing the response at the centre of the factorial design with the average of the responses at the corners. When interaction and curvature effects become appreciable compared with the experimental error (estimated by replication) a more elaborate factorial design is used which allows the form of the curved surface, and thus the approximate position of the maximum to be determined.

It is evident that factorial design and the method of steepest ascent will be very complicated when several factors are involved. The next section describes a method of optimization which is conceptually much simpler.

7.11 Simplex optimization

Simplex optimization may be applied when all the factors are continuous variables. A **simplex** is a geometrical figure which has $k + 1$ vertices when a response is being optimized with respect to k factors. For example, in the optimization of two factors the simplex will be a triangle. The method of optimization is illustrated by Figure 7.11. The initial simplex is defined by the points labelled 1, 2 and 3. In the first experiments the response is measured at each of the three combinations of factor levels given by the vertices of this triangle. The worst response in this case would be found at point 3 and it would be logical to conclude that a better response might be found at a point which is the reflection of 3 with respect to the line joining 1 and 2, i.e. at 4. The points 1, 2 and 4 form a new simplex and the response is measured for the combination of factor levels given by 4. (We immediately notice a

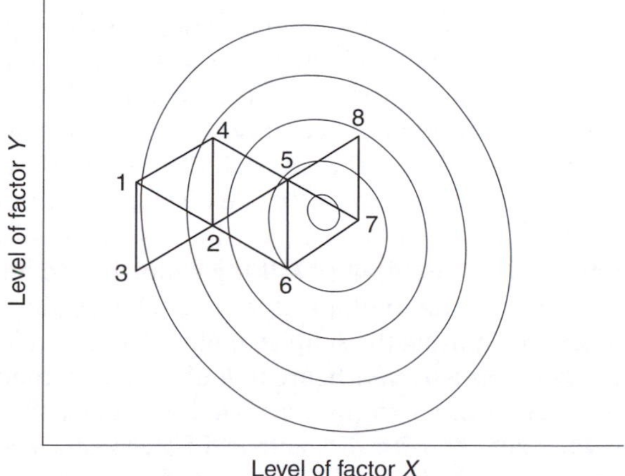

Figure 7.11 Simplex optimization.

major advantage of the simplex method, i.e. that at each stage of the optimization, only a single additional experiment is required.) Comparing the responses for the points 1, 2 and 4 will show that 1 now gives the worst response. The reflection process is repeated to give the simplex defined by 2, 4 and 5. The continuation of this process is shown in the figure. It can be seen that no further progress is possible beyond the stage shown, since points 6 and 8 both give a worse response than 5 and 7.

In order to improve the performance of the simplex method many modifications have been proposed. One obvious problem with the method is that, if the initial simplex is too small, too many experiments may be needed to approach the optimum. If the initial simplex is too big, the precision with which the optimum is determined will be poor. This problem can be overcome by advancing towards the optimum using a simplex that can vary in step size according to how the response for the new vertex in a simplex compares with the other vertices. Thus if vertex 4 in the above example gave a significantly better response than vertices 1–3, the new simplex might be stretched further in the same direction by moving vertex 4 twice as far from the line joining vertices 1 and 2. On the other hand, if the original vertex 4 gave a poorer response than vertex 1 then the simplex has probably been stretched too far, and the vertex 4 can be moved back so that it is only half as far from the line joining vertices 1 and 2. In other circumstances a 'negative' reflection might be appropriate, i.e. vertex 4 might fall *inside* the triangle formed by vertices 1–3. When two factors are optimized, the effect of these variable step sizes is that the triangles making up each simplex are not necessarily equilateral ones. The benefit of the variable step sizes is that initially the simplex is large, and gives rapid progress towards the optimum. Near the optimum it contracts to allow the latter to be found more accurately. When several factors are under study, it may be helpful to alter some of them by a constant step size, but others with a variable step size.

The position of the new vertex of a simplex is in practice found by calculation rather than drawing: this is essential when there are more than two factors. The calculation (using constant step sizes) is most easily set out as shown in Table 7.7, the calculation lines being labelled (i)–(v). In this example there are five factors and hence the simplex has six vertices. In the initial simplex the response for vertex 4 is the lowest and so this vertex is to be replaced. The co-ordinates of the centroid of the vertices which are to be *retained* are found by summing the co-ordinates for the retained vertices and dividing by the number of factors, k. The displacement of the new point from the centroid is given by (iv) = (ii) − (iii), and the co-ordinates of the new vertex, vertex 7, by (v) = (ii) + (iv). If the simplex is to be varied in size then the values in row (iv) are multiplied by a suitable scaling factor.

An obvious question in using the simplex method is the choice of the initial simplex. If this is taken as a *regular* figure in k dimensions, then the positions taken by the vertices in order to produce such a figure will depend on the scales used for the axes. As with the method of steepest ascent these scales should be chosen so that unit change in each factor gives roughly the same change in response. If there is insufficient information to achieve this, the

Table 7.7 Simplex optimization example

	Factors					Response
	A	*B*	*C*	*D*	*E*	
Vertex 1	1.0	3.0	2.0	6.0	5.0	7
Vertex 2	6.0	4.3	9.5	6.9	6.0	8
Vertex 3	2.5	11.5	9.5	6.9	6.0	10
Vertex 4 (rejected)	2.5	4.3	3.5	6.9	6.0	6
Vertex 5	2.5	4.3	9.5	9.7	6.0	11
Vertex 6	2.5	4.3	9.5	6.9	9.6	9
(i) Sum (excluding vertex 4)	14.50	27.40	40.00	36.40	32.60	
(ii) Sum/k (excluding vertex 4)	2.90	5.48	8.00	7.28	6.52	
(iii) Rejected vertex (i.e. 4)	2.50	4.30	3.50	6.90	6.00	
(iv) Displacement = (ii) − (iii)	0.40	1.18	4.50	0.38	0.52	
(v) Vertex 7 = (ii) + (iv)	3.30	6.66	12.50	7.66	7.04	

difference between the highest and lowest feasible value of each factor can be represented by the same distance. The size of the initial simplex is not so critical if it can be expanded or contracted as the method proceeds. Algorithms that can be used to calculate the initial positions of the vertices have been developed: one vertex is normally positioned at the currently accepted levels of the factors. This last point is a reminder that the analyst is rarely completely in the dark at the start of an optimization process. Previous experience will provide some guidance on feasible values for the vertices of the starting simplex.

It can be seen that in contrast to a factorial design the number of experiments required in the simplex method does not increase rapidly with the number of factors. For this reason all the factors which might reasonably be thought to have a bearing on the response should be included in the optimization.

Once an optimum has been found, the effect on the response when one factor is varied while the others are held at their optimum levels can be investigated for each factor in turn. This procedure can be used to check the optimization. It also indicates how important deviations from the optimum level are for each factor: the sharper the response peak in the region of the optimum the more critical any variation in factor level.

Simplex optimization has been used with success in many areas of analytical chemistry, e.g. atomic-absorption spectrometry, gas chromatography, colorimetric methods of analysis, plasma spectrometry, and the use of centrifugal analysers in clinical chemistry. When an instrument is interfaced with a microcomputer, the results of simplex optimization can be used to initiate automatic improvements in the instrument variables.

Occasionally response surfaces with more than one maximum occur, such as that shown in Figure 7.12. Both the alternating variable search and simplex

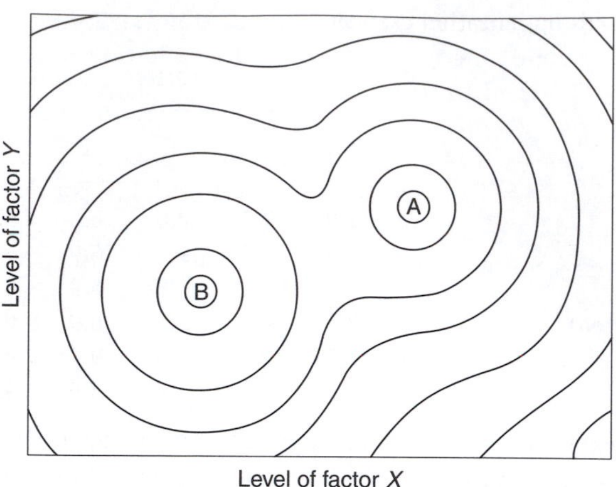

Figure 7.12 Contour diagram showing localized optimum (A) and true optimum B.

optimization methods may then locate a local optimum such as A rather than the true optimum B. Starting the optimization process in a second region of the factor space and verifying that the same optimum conditions are obtained is the preferred method for checking this point. Again the simplex method is valuable here, as it minimizes the extra work required.

7.12 Simulated annealing

In recent years there has been much interest in the application of calculation methods that mimic natural processes: these are collectively known as **natural computation** methods. Neural networks (see Chapter 8) are now being applied more frequently in analytical chemistry, and in the area of optimization **simulated annealing** (SA) has found some recent applications. Annealing is the process by which a molten metal or other material cools slowly in controlled conditions to a low energy state. In principle the whole system should be in equilibrium during the whole of the annealing process, but in practice random processes occur which result in short-lived and/or local *increases* in energy. When an analogous process is applied to an optimization problem the algorithm used allows access to positions in factor space that give a *poorer* response than the previous position. The result is that, unlike the AVS and simplex methods, which almost inevitably lead to the identification of an optimum which is closest to the starting point, SA methods can handle any local optima which occur, and successfully identify the true overall optimum.

In simple terms the method operates as follows. The first step is to identify, either at random or from experience, starting values for the levels of the k factors. These values give an initial response, R_1. In the second step a random vector, obtained using k random numbers, is added to the starting values, and a new set of experimental conditions generated: these yield a new response, R_2. As in other optimization methods, if R_2 is a better response than R_1, that is a good outcome, and the random addition step is repeated. The crucial characteristic of the method, however, is that *even if R_2 is a poorer response than R_1*, it is accepted as long as it is not much worse. (Clearly, numerical rules have to be applied to make this decision.) Eventually a situation arises in which a response is rejected, and, for example, five alternatives generated at random are also rejected as giving unacceptably poorer responses. In that situation it is assumed that the previous response was the optimum one. SA methods have been applied in UV-visible and near-IR spectroscopy, and in some cases found to be superior to simplex methods.

Bibliography

Gardner, W. P. 1997. *Statistical Analysis Methods for Chemists; A Software Based Approach*. Royal Society of Chemistry, Cambridge. (Extensive treatment of experimental designs in many branches of chemistry.)

Massart, D. L., Vandeginste, B. G. M., Buydens, L. M .C., De Jong, S., Lewi, P. J. and Smeyers-Verbeke, M. 1997. *Handbook of Chemometrics and Qualimetrics, Part A*. Elsevier, Amsterdam. (Good general coverage of optimization and experimental design.)

Morgan, E. 1991. *Chemometrics: Experimental Design*. Wiley, Chichester. (Clear introduction to factorial designs.)

Otto, M. 1999. *Chemometrics; Statistics and Computer Applications in Analytical Chemistry*. Wiley-VCH, Weinheim. (An excellent treatment of many topics covered in Chapters 7 and 8 of this book.)

Exercises

1. Four standard solutions were prepared, each containing 16.00% (by weight) of chloride. Three titration methods, each with a different technique of end-point determination, were used to analyse each standard solution. The order of the experiments was randomized. The results for the chloride found (% w/w) are shown below:

		Method	
Solution	*A*	*B*	*C*
1	16.03	16.13	16.09
2	16.05	16.13	16.15
3	16.02	15.94	16.12
4	16.12	15.97	16.10

Test whether there are significant differences between (a) the concentration of chloride in the different solutions, and (b) the results obtained by the different methods.

2. A new microwave assisted extraction method for the recovery of 2-chlorophenol from soil samples was evaluated by applying it to five different soils on each of three days. The percentage recoveries obtained were:

	Day		
Soil	*1*	*2*	*3*
1	67	69	82
2	78	66	76
3	78	73	75
4	70	69	87
5	69	71	80

Determine whether there were any significant differences in percentage recovery (a) between soils, and/or (b) between days.
(Data adapted from Egizabal, A., Zuloaga, O., Etxebarria, N., Fernández, L. A. and Madariaga, J. M. 1998. *Analyst* 123: 1679)

3. In studies of a fluorimetric method for the determination of the anionic surfactant sodium dodecyl sulphate (SDS) the interfering effects of four organic compounds at three different SDS : compound molar ratios were studied. The percentage recoveries of SDS were found to be:

	Molar ratios		
Organic compound	*1 : 1*	*1 : 2*	*1 : 3*
2,3-naphthalene dicarboxylic acid	91	84	83
Tannic acid	103	104	104
Phenol	95	90	94
Diphenylamine	119	162	222

Determine whether the SDS recovery depends on the presence of the organic compounds and/or on the molar ratios at which they are present. How should the experiment be modified to test whether any interaction effects are present?
(Recalde Ruiz, D. L., Carvalho Torres, A. L., Andrés Garcia, E. and Díaz García, M. E. 1998. *Analyst* 123: 2257)

4. Mercury is lost from solutions stored in polypropylene flasks by combination with traces of tin in the polymer. The absorbance of a standard aqueous solution of mercury stored in such flasks was measured for two levels of the following factors:

Factor	*Low*	*High*
A – Agitation of flask	Absent	Present
C – Cleaning of flask	Once	Twice
T – Time of standing	1 hour	18 hours

The following results were obtained. Calculate the main and interaction effects.

Combination of factor levels	*Absorbance*
1	0.099
a	0.084
c	0.097
t	0.076
ac	0.082
ta	0.049
tc	0.080
atc	0.051

(Adapted from Kuldvere, A., 1982. *Analyst* 107: 179)

5. In an interlaboratory collaborative experiment on the determination of arsenic in coal, samples of coal from three different regions were sent to each of three laboratories. Each laboratory performed a duplicate analysis on each sample, with the results shown below (measurements in $\mu g\ g^{-1}$).

	Laboratory		
Sample	*1*	*2*	*3*
A	5.1, 5.1	5.3, 5.4	5.3, 5.1
B	5.8, 5.4	5.4, 5.9	5.2, 5.5
C	6.5, 6.1	6.6, 6.7	6.5, 6.4

Verify that there is no significant sample–laboratory interaction, and test for significant differences between the laboratories.

6. The optimum pH for an enzyme-catalysed reaction is known to lie between 5 and 9. Determine the pH values at which the first two experiments of an optimization process should be performed in the following circumstances:

(a) The number of experiments and the degree of optimization needed are not pre-determined.
(b) The optimum pH needs to be known with a maximum range of 0.1 pH units.
(c) Only six experiments can be performed.

In (c) what is the degree of optimization obtained?

7. If the response at vertex 7 in the example on simplex optimization (pp. 208–9) is found to be 12, which vertex should be rejected in forming the new simplex and what are the co-ordinates of the new vertex?

CHAPTER EIGHT

Multivariate analysis

8.1 Introduction

Modern automatic analysis methods provide opportunities to collect large amounts of data very easily. For example, in clinical chemistry it is routine to determine many analytes for each specimen of blood, urine, etc. A number of chromatographic and spectroscopic methods can provide analytical data on many components of a single specimen. Situations like these, where several variables are measured for each specimen, yield **multivariate data**. One use of such data in analytical chemistry is in discrimination, for example determining whether an oil-spill comes from a particular source by analysing the fluorescence spectrum. Another use is classification, for example dividing the stationary phases used in gas–liquid chromatography into groups with similar properties by studying the retention behaviour of a variety of solutes with different chemical properties. In each case it would be possible to compare specimens by considering each variable in turn but modern computers allow more sophisticated processing methods where all the variables are considered simultaneously.

Each specimen, or, to generalize, each object is characterized by a set of measurements. When only two variables are measured this information can be represented graphically, as shown in Figure 8.1, where the co-ordinates of the point give the values taken by the two variables. The point can also be defined by a vector, called a **data vector**, drawn to it from the origin. Objects which have similar properties will have similar data vectors, that is they will lie close to each other in the space defined by the variables. Such a group is called a **cluster**.

A graphical representation is less easy for three variables and no longer possible for four or more: it is here that computer analysis is particularly valuable in finding patterns and relationships. Matrix algebra is needed

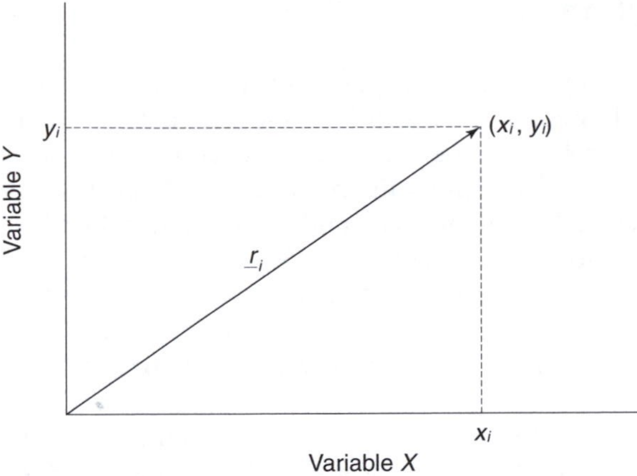

Figure 8.1 A diagram to illustrate a data vector, $\underline{r}_i$. x_i and y_i are the values taken by the variables X and Y respectively.

in order to describe the methods of multivariate analysis fully. No attempt will be made to do this here. The aim is to give an appreciation of the purpose and power of multivariate methods. Simple data sets will be used to illustrate the methods and some practical applications will be described.

Table 8.1 The intensity of the fluorescence spectrum at four different wavelengths for a number of compounds

	Wavelength (nm)			
Compound	*300*	*350*	*400*	*450*
A	16	62	67	27
B	15	60	69	31
C	14	59	68	31
D	15	61	71	31
E	14	60	70	30
F	14	59	69	30
G	17	63	68	29
H	16	62	69	28
I	15	60	72	30
J	17	63	69	27
K	18	62	68	28
L	18	64	67	29
Mean	15.75	61.25	68.92	29.25
Standard deviation	1.485	1.658	1.505	1.485

8.2 Initial analysis

Table 8.1 shows an example of some multivariate data. This gives the relative intensities of fluorescence emission at four different wavelengths (300, 350, 400, 450 nm) for 12 compounds, A–L. In each case the emission intensity at the wavelength of maximum fluorescence would be 100. As a first step it may be useful to calculate the mean and standard deviation for each variable. These are also shown in the table.

In addition, since we have more than one variable, it is possible to calculate a product–moment (Pearson) correlation coefficient for each pair of variables. These are summarized in the **correlation matrix** in Table 8.2, obtained using Minitab.

This shows that, for example, the correlation coefficient for the intensities at 300 and 350 nm is 0.914. The relationships between pairs of variables can be illustrated by a **draftsman plot** as shown in Figure 8.2. This gives scatter diagrams for each pair of variables. Both the correlation matrix and the scatter diagrams indicate that there is some correlation between some of the pairs of variables.

Table 8.2 The correlation matrix for the data in Table 8.1

Correlations (Pearson)

	300	350	400
350	0.914		
400	-0.498	-0.464	
450	-0.670	-0.692	0.458

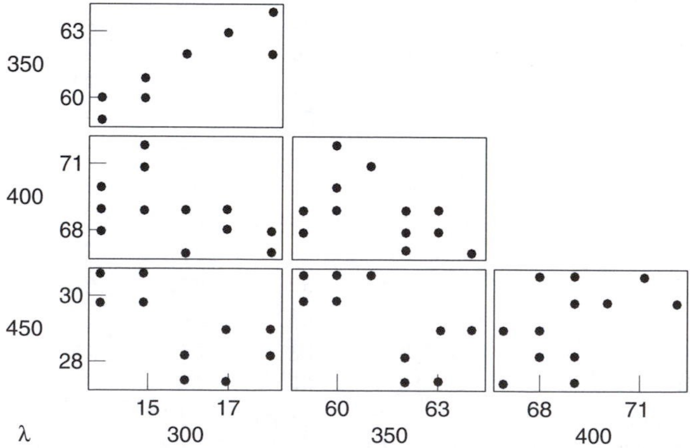

Figure 8.2 Draftsman plot for the data in Table 8.1.

8.3 Principal component analysis

One problem with multivariate data is that its sheer volume may make it difficult to see patterns and relationships. For example, a spectrum would normally be characterized by several hundred intensity measurements rather than just four as in Table 8.1 and in this case the correlation matrix would contain hundreds of values. Thus the aim of many methods of multivariate analysis is data reduction. Quite frequently there is some correlation between the variables, as there is for the data in Table 8.1, and so some of the information is redundant. **Principal component analysis** (PCA) is a technique for reducing the amount of data when there is correlation present. It is worth stressing that it is not a useful technique if the variables are uncorrelated.

The idea behind PCA is to find **principal components** $Z_1, Z_2, \ldots, Z_n$ which are linear combinations of the original variables describing each specimen, $X_1, X_2, \ldots, X_n$, i.e.

$$Z_1 = a_{11}X_1 + a_{12}X_2 + a_{13}X_3 + \cdots a_{1n}X_n$$

$$Z_2 = a_{21}X_1 + a_{22}X_2 + a_{23}X_3 + \cdots a_{2n}X_n$$

etc.

For example, for the data in Table 8.1 there would be four principal components Z_1, Z_2, Z_3 and Z_4, each of which would be a linear combination of X_1, X_2, X_3 and X_4, the fluorescence intensities at the given wavelengths. The coefficients, a_{11}, a_{12}, etc. are chosen so that the new variables, unlike the original variables, are not correlated with each other. Creating a new set of variables in this way may seem a pointless exercise since we obtain n new variables in place of the n original ones, and hence no reduction in the amount of data. However, the principal components are also chosen so that the first principal component (PC1), Z_1, accounts for most of the variation in the data set, the second (PC2), Z_2, accounts for the next largest variation and so on. Hence, when significant correlation occurs the number of useful PCs is much less than the number of original variables.

Figure 8.3 illustrates the method when there are only two variables and hence only two principal components. In Figure 8.3a the principal components are shown by the dotted lines. The principal components are at right-angles to each other, a property known as **orthogonality**. Figure 8.3b shows the points referred to these two new axes and also the projection of the points on to PC1 and PC2. We can see that in this example Z_1 accounts for most of the variation and so it would be possible to reduce the amount of data to be handled by working in one dimension with Z_1 rather than in two dimensions with X_1 and X_2. (In practice we would not need to use PCA when there are only two variables because such data are relatively easy to handle.)

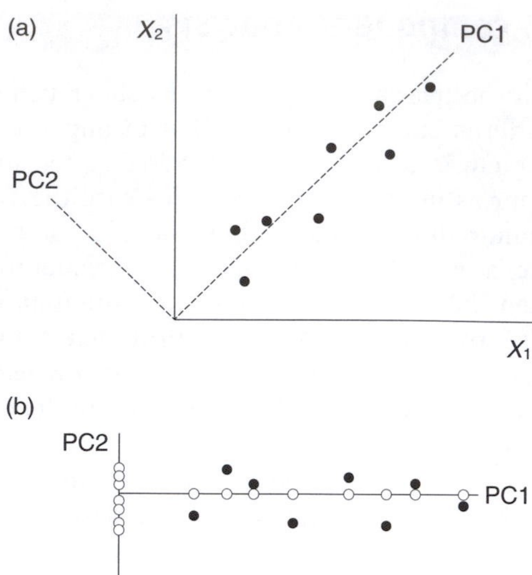

Figure 8.3 (a) Diagram illustrating the two principal components, PC1 and PC2, for the two variables, X_1 and X_2. (b) Points referred to the principal component axes. • indicates data points, ○ their projection onto the axes.

Figure 8.3 shows that PCA is equivalent to a rotation of the original axes in such a way that PC1 is in the direction of maximum variation, but with the angle between the axes unchanged. With more than two variables it is not possible to illustrate the method diagrammatically but again we can think of PCA as a rotation of the axes in such a way that PC1 is in the direction of maximum variation, PC2 is in the direction of next greatest variation, and so on. It is often found that PC1 and PC2 then account between them for most of the variation in the data set. As a result the data can be represented in only two dimensions instead of the original n.

One decision which must be made before carrying out a PCA is whether to use raw data or to first standardize each variable to zero mean and unit variance. If the variables are not standardized and one variable has a much larger variance, then this variable will dominate the first principal component. Standardizing avoids this by making all the variables carry equal weight: in the discussion which follows we will assume that the data have been standardized.

In mathematical terms the principal components are the eigenvectors of the correlation matrix and the technique for finding these eigenvectors is called eigenanalysis. Corresponding to each principal component (i.e. eigenvector) is an eigenvalue which gives the amount of variance in the data set which is explained by that principal component. For standardized data each of the original variables has a standard deviation, and hence a variance, of 1. Thus the total variance of the data set and the sum of the eigenvalues are both equal to the number of variables.

EXAMPLE 8.3.1

Carry out a principal component analysis of the data in Table 8.1.

This can be done using a variety of computer packages (for example Minitab, SAS, The Unscrambler..). The printout below was obtained from Minitab.

Principal Component Analysis

```
Eigenanalysis of the Correlation Matrix

Eigenvalue        2.8807      0.6453      0.3897      0.0844
Proportion         0.720       0.161       0.097       0.021
Cumulative         0.720       0.881       0.979       1.000

Variable             PC1         PC2         PC3         PC4
300                0.547      -0.238      -0.395       0.699
350                0.546      -0.299      -0.324      -0.712
400               -0.400      -0.913       0.073       0.043
450               -0.493       0.145      -0.856      -0.049
```

In this example there are four variables and so the sum of the variances for the (now standardized) data is 4. The first line labelled 'eigenvalue' shows how this variance is shared between the four principal components with PC1 having a variance of 2.8807, PC2 having a variance of 0.6453, and so on. Note that, as we would expect, PC1 has the highest variance, PC2 the next highest and so on. Principal components with an eigenvalue greater than 1 contribute more to the variance than the original variables. The sum of the variances of the four principal components is 4 (allowing for rounding errors). Again this is what we would expect since the principal components between them must explain all the variation in the data set. The second line of the table (labelled 'proportion') gives the proportion of the variation explained by each principal component. The line below gives the cumulative proportion. It shows, for example, that between them PC1 and PC2 accounted for 88.1% of the variation.

The bottom half of the table gives the coefficients of the principal components. (The coefficients are scaled so that the length of a data vector is unaltered by the change in variables. The sign of the first coefficient of the first principal component is (arbitrarily) chosen to be positive.) For example, the first principal component is $Z_1 = 0.547X_1 + 0.546X_2 - 0.400X_3 - 0.493X_4$ where X_1, X_2, X_3 and X_4 are the standardized relative intensities at 300, 350, 400 and 450 nm respectively. Each original variable is standardized by subtracting the mean for that variable and then dividing by the corresponding standard deviation (obtained from Table 8.1). Thus for compound A,

$$\begin{aligned} Z_1 &= 0.547 \times \frac{(16 - 15.75)}{1.485} + 0.546 \times \frac{(62 - 61.25)}{1.658} - 0.400 \times \frac{(67 - 68.92)}{1.505} \\ &\quad - 0.493 \times \frac{(27 - 29.25)}{1.485} \\ &= 1.60 \end{aligned}$$

This value is sometimes referred to as a 'score' for PC1. Figure 8.4 plots the scores of the first two principal components, calculated in this way, for the compounds A–L. This diagram reveals that the compounds fall into two distinct groups, a fact which is not readily apparent from the original data.

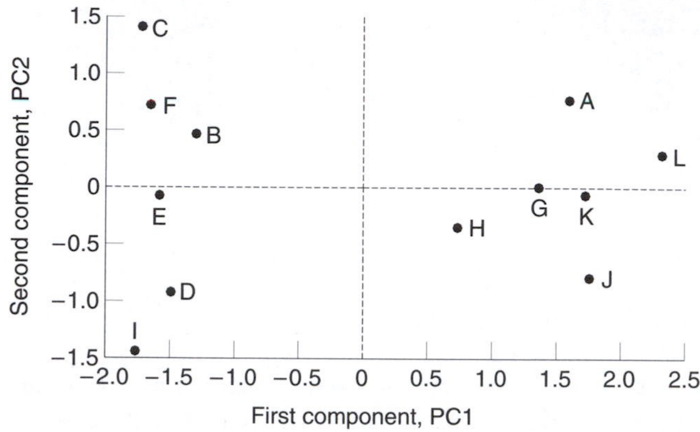

Figure 8.4 The scores of the first two principal components for the data in Table 8.1.

Table 8.3 shows the original data rearranged so that compounds with similar spectra are grouped together. The differences between the two groups are now apparent. There is a difference at all four wavelengths, and the magnitudes of these differences are similar. This corresponds to the fact that the coefficients for the first principal component are similar in size. The top group in Table 8.3 has higher intensities than the bottom group at 300 and 350 nm and the opposite is true at 400 and 450 nm. This corresponds to the fact the first two coefficients of Z_1 have the opposite sign from the second two. Once two or more groups have been identified by using PCA, it may be possible to explain the differences between them in terms of chemical structure. Sometimes it may be possible to give a physical interpretation to the principal components. For this reason, principal components are sometimes referred to as **latent** (i.e. hidden) **variables**.

In this example the values of the coefficients show that each of the variables contributes to PC1 and at least three of them contribute to PC2. In other cases it is found that some variables do not contribute significantly even to PC1. An important benefit of PCA is that such variables can then be rejected.

PCA is primarily a mathematical method for data reduction and it does not assume that the data have any particular distribution. We have seen how PCA can be used to reduce the dimensionality of a data set and how it may thus reveal clusters. It has been used, for example, on the results of Fourier transform spectroscopy in order to reveal differences between hair from different racial groups and for classifying different types of cotton fibre. In another

Table 8.3 The data in Table 8.1 rearranged so that compounds with similar spectra are grouped together

	Wavelength (nm)			
Compound	*300*	*350*	*400*	*450*
A	16	62	67	27
G	17	63	68	29
H	16	62	69	28
J	17	63	69	27
K	18	62	68	28
L	18	64	67	29
B	15	60	69	31
C	14	59	68	31
D	15	61	71	31
E	14	60	70	30
F	14	59	69	30
I	15	60	72	30

example the concentrations of a number of chlorobiphenyls were measured in specimens from a variety of sea mammals. A PCA of the results revealed differences between species, differences between males and females, and differences between young and adult individuals. PCA also finds application in multiple regression (see Section 8.8).

8.4 Cluster analysis

Although PCA may reveal groups of like objects, it is not always successful in doing so. Figure 8.5 shows a situation in which the first principal component

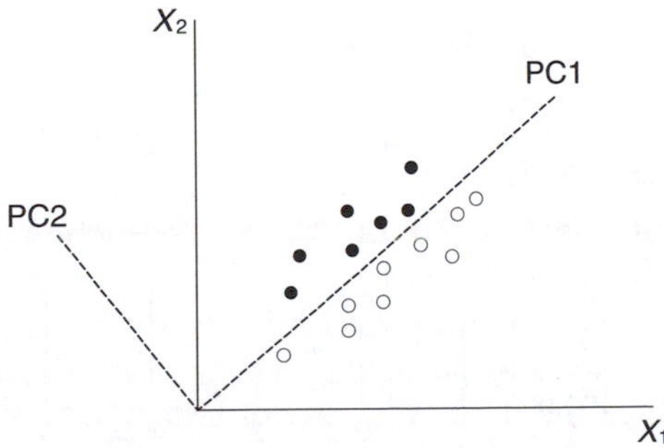

Figure 8.5 A situation in which the first principal component does not give a good separation between two groups.

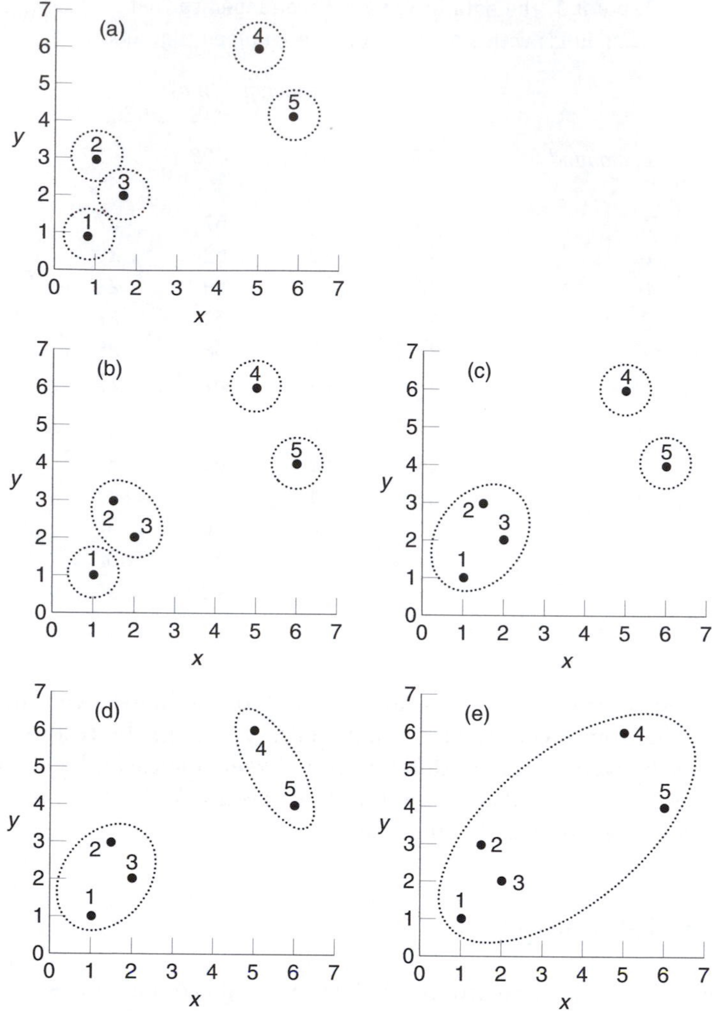

Figure 8.6 Stages in clustering: the dotted lines enclose clusters.

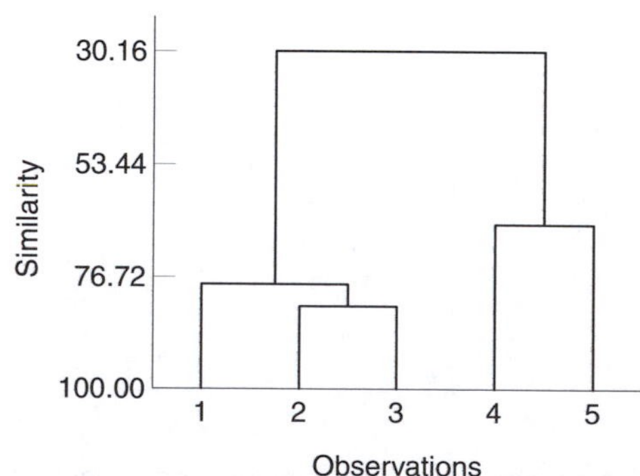

Figure 8.7 A dendrogram illustrating the stages of clustering for Figure 8.6.

does not give a good separation between two groups. In this section we turn to methods whose explicit purpose is to search for groups.

Cluster analysis is a method for dividing a group of objects into classes so that similar objects are in the same class. As in PCA, the groups are not known prior to the mathematical analysis and no assumptions are made about the distribution of the variables. Cluster analysis searches for objects which are close together in the variable space. The distance, d, between two points in n-dimensional space with co-ordinates $(x_1,x_2,\ldots,x_n)$ and $(y_1,y_2,\ldots,y_n)$ is usually taken as the **Euclidian distance** defined by

$$d = \sqrt{(x_1 - y_1)^2 + (x_2 - y_2)^2 + \cdots + (x_n - y_n)^2}$$

For example the distance between the compounds E and F in Table 8.3 (if the unstandardized variables are used) is given by:

$$d = \sqrt{(14 - 14)^2 + (60 - 59)^2 + (70 - 69)^2 + (30 - 30)^2} = \sqrt{2}$$

As in PCA, a decision has to made as to whether or not the data are standardized. Standardizing the data will mean that all the variables are measured on a common scale so that one variable does not dominate the others.

There are a number of methods for searching for clusters. One method starts by considering each object as forming a 'cluster' of size one, and compares the distances between these clusters. The two points which are closest together are joined to form a new cluster. The distances between the clusters are again compared and the two nearest clusters combined. This procedure is repeated and, if continued indefinitely, will group all the points together. There are a variety of ways of computing the distance between two clusters which contain more than one member. The simplest conceptually is to take the distance between two clusters as the distance between nearest neighbours. This is called the **single linkage method**. It is illustrated in Figure 8.6. The successive stages of grouping can be shown on a **dendrogram** as in Figure 8.7. The vertical axis can either show the distance, d_{ij}, between two points i and j when they are joined or alternatively the **similarity**, s_{ij}, defined by $s_{ij} = 100(1 - d_{ij}/d_{\max})$ where $d_{\max}$ is the maximum separation between any two points. The resulting diagrams look the same but their vertical scales differ. The stage at which the grouping is stopped, which determines the number of clusters in the final classification, is a matter of judgement for the person carrying out the analysis.

EXAMPLE 8.4.1

Apply the single linkage method to the (unstandardized) data in Table 8.1.

The printout below was obtained using Minitab. With this software the linkages continue until there is only one cluster unless the user specifies otherwise.

Hierarchical Cluster Analysis of Observations

Euclidean Distance, Single Linkage

Amalgamation Steps

Step	Number of clusters	Similarity level	Distance level	Clusters joined		New cluster	Number of Obs in new cluster
1	11	80.20	1.414	5	6	5	2
2	10	80.20	1.414	3	5	3	3
3	9	75.75	1.732	7	12	7	2
4	8	75.75	1.732	7	11	7	3
5	7	75.75	1.732	8	10	8	2
6	6	75.75	1.732	4	9	4	2
7	5	75.75	1.732	2	3	2	4
8	4	71.99	2.000	7	8	7	5
9	3	71.99	2.000	2	4	2	6
10	2	68.69	2.236	1	7	1	6
11	1	49.51	3.606	1	2	1	12

The dendrogram in Figure 8 8 illustrates the stages of the linkage. The vertical scale gives the distance between the two groups at the point when they were combined. The table above shows that the first two points to be joined were 5 (compound E) and 6 (compound F) with a separation of 1.414 ($= \sqrt{2}$ as calculated earlier). The reader can verify that the distance of C from F is also $\sqrt{2}$ so the next stage is to join point 3 to the cluster consisting of points 5 and 6. The process continues until all the points are in one cluster. However, if we 'cut the tree' i.e. stop the grouping, at the point indicated by the dotted line in Figure 8.8, this analysis suggests that the compounds A–L fall into two distinct groups. Not surprisingly, the groups contain the same members as they did with PCA.

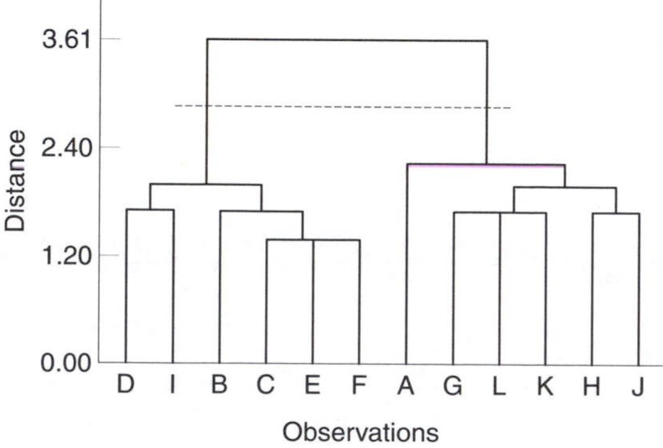

Figure 8.8 A dendrogram for the data in Table 8.1.

The method of cluster analysis described here is **hierarchical**, meaning that once an object has been assigned to a group the process cannot be reversed. For non-hierarchical methods the opposite is the case. One such method is the ***k*-means method** which is available, for example, in Minitab. This starts by either dividing the points into k clusters or alternatively choosing k 'seed points'. Then each individual is assigned to the cluster (or seed point) whose centroid is nearest. When a cluster loses or gains a point the position of the centroid is recalculated. The process is continued until each point is in the cluster whose centroid is nearest.

This method has the disadvantage that the final grouping reflects the initial choice of clusters or seed points. Another disadvantage is that the value of k has to be chosen in advance. Many methods have been suggested for deciding on the best value of k but none of them is really satisfactory.

Cluster analysis has been used to classify the many phases used in gas–liquid chromatography. A small preferred set of phases can then be selected by taking one phase from each cluster: this provides a range of stationary phases, each with distinctive separation characteristics. Another application is the classification of antibiotics in terms of their activity against various types of bacteria in order to elucidate the relationship between biological activity and molecular structure. A further recent application of cluster analysis is the classification of wine vinegars on the basis of a variety of organic and inorganic constituents.

8.5 Discriminant analysis

The methods described so far in this chapter have helped us to see whether objects fall into groups when we have no prior knowledge of the groups to be expected. Such methods are sometimes called **unsupervised pattern recognition**. We will now turn to so-called **supervised pattern recognition**. Here we start with a number of objects whose group membership is known, for example apple juices extracted from different varieties of fruit. These objects are sometimes called the **learning** or **training objects**. The aim of supervised pattern recognition methods is to use these objects to find a rule for allocating a new object of unknown group to the correct group.

The starting point of **linear discriminant analysis** (LDA) is to find a **linear discriminant function** (LDF), Y, which is a linear combination of the original variables X_1, X_2, etc:

$$Y = a_1X_1 + a_2X_2 + \cdots a_nX_n$$

The original n measurements for each object are combined into a single value of Y, so the data have been reduced from n dimensions to one dimension. The coefficients of the terms are chosen in such a way that Y reflects the *difference* between groups as much as possible: objects in the same group will have similar values of Y and objects in different groups will have very different values of Y. Thus the linear discriminant function (LDF) provides a means of discriminating between the two groups.

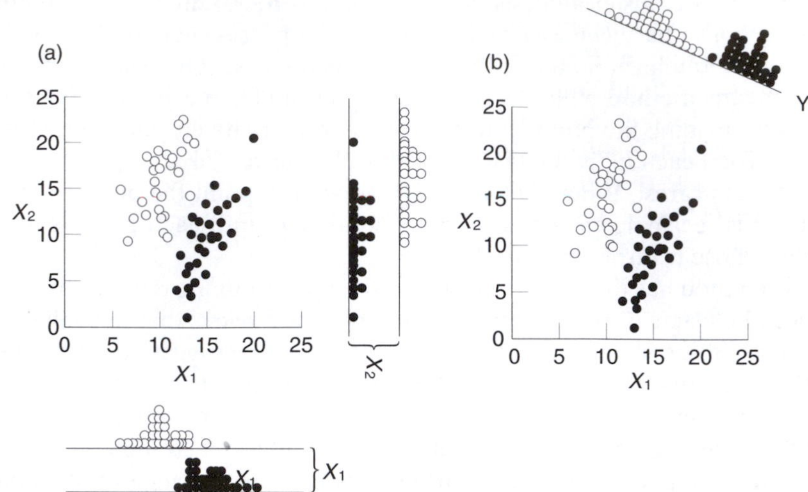

Figure 8.9 (a) Two groups and the distributions of each variable for each group. (b) The distribution of the linear discriminant function for each group.

The simplest situation is that in which there are two classes and two variables, X_1 and X_2, as illustrated in Figure 8.9a. This diagram also shows the distribution of the individual variables for each group in the form of dot-plots. For both the variables, there is a considerable overlap in the distributions for the two groups. It can be shown that the LDF for these data is $Y = 0.91X_1 + 0.42X_2$. This LDF is shown by the line labelled Y in Figure 8.9b and the value which the function takes for a given point is given by the projection of the point on to this line. Figure 8.9b shows the dot-plots of the LDF, Y, for each group. It can be seen that there is no overlap between the distribution of Y for the two groups. This means that Y is better at discriminating between the groups than the original variables.

An unknown object will be classified according to its Y value. A initial common sense approach would be to compare Y with $\bar{Y}_1$ and $\bar{Y}_2$, the Y values for the means of the two groups. If Y is closer to $\bar{Y}_1$ than to $\bar{Y}_2$ then the object belongs to group 1, otherwise it belongs to group 2. For these data, $\bar{Y}_1 = 3.15$ and $\bar{Y}_2 = 10.85$. So if $Y - 3.15 < 10.85 - Y$, that is $Y < 7.0$ we classify the object in group 1, otherwise we classify it in group 2. This method is satisfactory only if the two groups have similarly shaped distributions. Also, if experience shows that a single object is more likely to belong to one of the groups rather than the other, then the decision rule will need to be modified. Software such as Minitab permits such a modification.

The success of LDA at allocating an object correctly can be tested in several ways. The simplest is to use the classification rule to classify each object in the group and to record whether the resulting classification is correct. The table summarizing the results of this procedure is sometimes called the **confusion matrix** (always displayed in Minitab). This method tends to be over-optimistic since the object being classified was part of the set which was used to form the rule. A better method divides the original data into two groups chosen at random. The first group, known as the **training set**, is used to find the LDF. Then the objects in the second group (the **test set**) are allocated using this function and a success rate found. A third method, which uses the data more economically, is **cross-validation**, sometimes called the 'leave one out method'. As the latter name suggests, this finds the LDF with one object omitted and checks whether this LDF then allocates the omitted object correctly. The procedure is then repeated for each object in turn and again a success rate can be found. This method is an option in Minitab.

If the distributions do not have similar shapes, then a modification of LDA, known as **quadratic discriminant analysis** (QDA) may be used. This method assumes that the two groups have multivariate normal distributions but with different variances.

LDA and QDA can both be extended to the situation where there are more than two groups of objects. To avoid complex decision rules of the type given above (if $y - 3.15 < 10.85 - y$, etc.) many programs assume a multivariate normal distribution and find a new function, which includes a constant term, for each group. From these functions a score is calculated for each new object and the object is assigned to the group for which the score is highest. This is illustrated in the following example.

EXAMPLE 8.5.1

The table below gives the concentration in g l^{-1} of sucrose, glucose, fructose and sorbitol in apple juice from three different sources, A, B and C. Carry out an LDA and evaluate the method using cross-validation.

Variety	*Sucrose*	*Glucose*	*Fructose*	*Sorbitol*
A	20	6	40	4.3
A	27	11	49	2.9
A	26	10	47	2.5
A	34	5	47	2.9
A	29	16	40	7.2
B	6	26	49	3.8
B	10	22	47	3.5
B	14	21	51	6.3
B	10	20	49	3.2
B	8	19	49	3.5
C	8	17	55	5.3
C	7	21	59	3.3
C	15	20	68	4.9
C	14	19	74	5.6
C	9	15	57	5.4

Classify an apple juice with 11, 23, 50 and 3.9 g l^{-1} of sucrose, glucose, fructose and sorbitol, respectively.

The analysis below was obtained using Minitab.

```
Discriminant Analysis

Linear Method for Response: Variety
Predictors: Sucrose Glucose Fructose Sorbitol

Group        A        B        C
Count        5        5        5

Summary of Classification

Put into          ....True Group....

Group                 A         B         C

A                     5         0         0
B                     0         5         0
C                     0         0         5
Total N               5         5         5
N Correct             5         5         5
Proportion        1.000     1.000     1.000

N = 15       N Correct = 15      Proportion Correct = 1.000

Summary of Classification with Cross-validation

Put into          ....True Group....

Group                 A         B         C

A                     5         0         0
B                     0         5         0
C                     0         0         5
Total N               5         5         5
N Correct             5         5         5
Proportion        1.000     1.000     1.000

N = 15       N Correct = 15      Proportion Correct = 1.000

Linear Discriminant Function for Group

                      A         B         C

Constant         -44.19    -74.24   -114.01
Sucrose            0.39     -1.66     -2.50
Glucose            0.42      1.21      0.54
Fructose           1.46      2.53      3.48
Sorbitol           2.19      3.59      5.48
```

The 'summary of classification' gives the confusion matrix and shows a 100% success rate. The 'summary of classification with cross-validation' also shows a 100% success rate.

For the new apple juice the linear discriminant scores for each group have values:

Group A: $-44.19 + 0.39 \times 11 + 0.42 \times 23 + 1.46 \times 50 + 2.19 \times 3.9 = 51.301$
Group B: $-74.24 - 1.66 \times 11 + 1.21 \times 23 + 2.53 \times 50 + 3.59 \times 3.9 = 75.831$
Group C: $-114.01 - 2.5 \times 11 + 0.54 \times 23 + 3.48 \times 50 + 5.48 \times 3.9 = 66.282$

The score for group B is highest, so the unknown apple juice is presumed to have come from source B.

Unlike the other procedures described in this chapter, standardizing the variables has no effect on the outcome of linear discriminant analysis: it merely re-scales the axes. It may, however, be useful to work with standardized variables in order to decide which variables are important in providing discrimination between the groups. As a general guide it will be those variables which have the larger coefficients in the linear discriminant functions. Once these variables have been identified, the performance of the method with fewer variables can be investigated to see whether a satisfactory discrimination between the groups can still be achieved (see Exercise 1 at the end of this chapter).

Some recent applications of LDA include the classification of vegetable oils using the data obtained from an array of gas sensors and the use of proton magnetic resonance spectra to discriminate between normal and cancerous ovarian tissue.

Although the above method appears to analyse all the groups simultaneously, the method is actually equivalent to analysing the groups pairwise. An alternative method for more than two groups which genuinely analyses them simultaneously is **canonical variate analysis** (CVA). This is an extension of LDA which finds a number of **canonical variates** Y_1, Y_2, etc. (which are again linear combinations of the original variables). As with LDA, Y_1 is chosen in such a way that it reflects the difference between the groups as much as possible. Then Y_2 is chosen so that it reflects as much of the remaining difference between the groups as possible, subject to the constraint that there is no correlation between Y_1 and Y_2, and so on. CVA could be thought of as PCA for groups but, unlike PCA, the results are not dependent on scale, so no pre-treatment of the data is necessary.

The following section describes an alternative method which can be used when there are two or more groups.

8.6 *K*-nearest neighbour method

This is a conceptually simple method for deciding how to classify an unknown object when there are two or more groups of objects of known class. It makes no assumption about the distribution in the classes and can be used when the groups cannot be separated by a plane, as illustrated in Figure 8.10. In its simplest form an unknown object is allocated to the class of its nearest neighbour. Alternatively, the K nearest neighbours (where K is a small integer) are

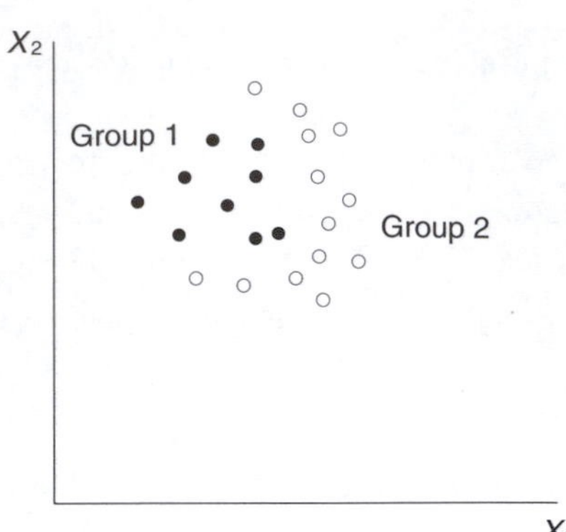

Figure 8.10 Two groups which cannot be separated by a plane.

taken and the class membership is decided by a voting scheme. For example, provided K is odd, the unknown object can be allocated to the class of the majority of its K nearest neighbours. In more sophisticated versions, different weightings can be given to the neighbours, depending on their relative distances.

8.7 Disjoint class modelling

The emphasis in the methods described in Sections 8.5 and 8.6 has been on trying to find a boundary between two or more classes, so that an unknown object may be allocated to the correct class. However, the situation may arise when the unknown object does not belong to any of the classes being considered. For example, in Example 8.5.1, it was assumed that the unknown apple juice came from one of the sources A, B or C. However it might have come from none of these sources but we would have still (incorrectly) allocated it to one of them. A different approach is needed if this sort of error is to be avoided. Instead of having a rule which discriminates between classes, we need a rule which allows us to discriminate between membership and non-membership of a given class. This is done by making a separate model for each class and using the model in order to test whether the unknown object could be a member of the class. This is called **disjoint class modelling**. For example, if the number of variables is small, each class might be modelled by a multivariate normal distribution. With more variables, some data reduction needs to be carried out first. One such method, called **SIMCA (Soft Independent Modelling of Class Analogy)**, makes a model of each class in terms of the first few principal components for that class.

8.8 Multiple regression

We turn now to the situation in which our variables can be divided into two groups: response variables and predictor variables. The situation in which we have one response variable, y, depending on a number of predictor variables,

x_1, x_2, x_3, etc., is known as **multiple regression**. A simple example would be the situation where y is an absorbance value from a mixture of compounds with concentrations $x_1, x_2, x_3 \ldots$. The techniques of linear regression, which were described in Chapter 5, can be extended to find a regression equation in the form

$$y = b_0 + b_1x_1 + b_2x_2 + \cdots$$

In order to carry out multiple regression the number of calibration specimens must be greater than the number of predictors. This is likely to be the case in the situation given above but may not always be so as we shall see in Section 8.11.

As with univariate regression, an analysis of the residuals is important in evaluating the model. The residuals should be randomly and normally distributed. The prediction performance can be validated in a way similar to the validation of LDA, i.e. either by dividing the data into two randomly chosen groups, making the model with one group and then testing it with the other, or by using a 'leave-one-out' method. A graph of the predicted values against the measured values will give points close to a straight line if the model is a satisfactory one.

Unlike the univariate situation, there is the option of omitting some of the predictor variables. We could, if we wished, try all possible combinations of the predictor variables and find the one which predicts y successfully with the minimum number of predictor variables. The adjusted value of R^2 (see Section 5.14) can be used to compare the performance of the different models.

8.9 Principal component regression

One problem in multiple regression is that correlations between the predictor variables can lead to mathematical complications, resulting in unreliable predictions of y. A way round this is to carry out a PCA on the x variables and then to regress y on the principal components. This is known as **principal component regression** (PCR). Since the principal components are not correlated (see Section 8.3) the problem of correlation between the predictor variables is overcome.

PCR is also a valuable technique when the number of original predictor variables exceeds the number of calibration specimens available. The number of predictor variables can be reduced by using the first few principal components rather than the original variables. This method will give satisfactory results provided that the principal components used account between them for most of the variation in the predictor variables. This technique is often used in multivariate calibration (see Section 8.12).

8.10 Multivariate regression

The term **multivariate regression** is usually applied to the situation in which there is a multivariate response. If there is one predictor variable a possible method of analysis would be to find a regression equation relating each of the response variables, y_i, to the predictor variable. If there is more than one predictor we could carry out a multiple regression of each y_i on the predictor variables. Alternatively, we could first do a PCA on the predictor variables to produce new uncorrelated variables and then carry out a multiple regression of each y_i on these principal components. Another possibility would be to find principal components of the response variables and regress them on principal components of the predictor variables.

The next section describes a method which utilizes the correlations between the response and predictor variables rather than applying PCA approaches to the two groups of variables separately.

8.11 Partial least squares regression

Like PCR, **partial least squares regression** (PLS regression) starts by finding linear combinations of the predictor variables. However the way in which these linear combinations are chosen is different. In PCR the principal components are chosen so that they describe as much of the variation in the predictors as possible, irrespective of the strength of the relationships between the predictor and response variables. In PLS, variables which show a high correlation with the response variables are given extra weight because they will be more effective at prediction. In this way linear combinations of the predictor variables are chosen which are highly correlated with the response variables and also explain the variation in the predictor variables. A distinction is usually made between the situation when the response consists of a single variable and that when the response is multivariate: the former is called PLS1, the latter PLS2.

8.12 Multivariate calibration

As noted above, an example of the application of multivariate regression is in the determination of the concentration of the constituents in a mixture of analytes by spectral analysis. In the classical approach the intensity, y_i, at each of a number of wavelengths would be related to the concentrations of the constituents by an equation of the form $y_i = b_{0i} + b_{1i}x_1 + b_{2i}x_2 + \cdots$ where the coefficients for each constituent are dependent on wavelength. Then, from the measured spectrum of a specimen with unknown composition (i.e. a test specimen), the concentrations of the analytes in this specimen could be estimated. This method is the multivariate analogue of the univariate method described in Section 5.4. The method requires knowledge of the spectra of the pure constituents and calibration specimens of known composition. It assumes that there are no other components in the specimens which interfere with the components of interest, in the concentration range used, and that components of interest do not interfere with each other.

In many cases the prediction specimens do contain substances other than those of interest and these act as interferents. If this is the case, it is better to calibrate with specimens from similar sources (which will have similar composition) and to use **inverse calibration**. This means that the analyte concentration is modelled as a function of the spectrum (i.e. the reverse of the classical method). Inverse calibration is appropriate since the concentration is no longer a controlled variable. Even when it is possible to prepare calibration specimens of known composition, if the substances of interest interfere with each other then the concentrations of these substances are in effect no longer controlled variables: in these circumstances inverse calibration is again appropriate. The following example illustrates the method.

EXAMPLE 8.12.1

The table below gives the UV absorbance (×100) recorded at three different wavelengths, A_1, A_2 and A_3, of 10 specimens (A–J) and the measured concentrations (mM), c_1, c_2, c_3 and c_4 of four constituents of interest.

Specimen	c_1	c_2	c_3	c_4	A_1	A_2	A_3
A	0.888	0.016	0.014	0.082	91.5	56.1	73.6
B	0.461	0.091	0.243	0.205	93.8	56.3	74.1
C	0.453	0.159	0.233	0.156	93.4	56.4	74.5
D	0.560	0.093	0.085	0.263	92.5	56.7	73.7
E	0.414	0.019	0.279	0.289	94.8	56.5	73.6
F	0.438	0.169	0.137	0.256	93.2	56.8	73.9
G	0.342	0.228	0.196	0.233	93.7	57.0	74.4
H	0.743	0.109	0.006	0.142	91.5	56.8	73.9
I	0.751	0.011	0.148	0.090	92.7	55.7	73.9
J	0.477	0.146	0.063	0.314	92.7	57.7	73.8

Find the regression equation for predicting c_1, c_2, c_3 and c_4 from A_1, A_2 and A_3.

The print-out below was obtained using Minitab and gives the regression equation for c_1.

```
Regression Analysis

The regression equation is
c1 = 31.7 - 0.129 A1 - 0.153A2 - 0.142 A3

Predictor     Coef          StDev         T          P
Constant    31.688          3.999      7.92      0.000
A1        -0.12893        0.01576     -8.18      0.000
A2        -0.15260        0.02863     -5.33      0.002
A3        -0.14214        0.05228     -2.72      0.035

S = 0.04664      R-Sq = 95.5%      R-Sq(adj) = 93.2%
```

```
Analysis of Variance

Source          DF         SS          MS         F        P
Regression       3    0.275833    0.091944    42.27    0.000
Error            6    0.013051    0.002175
Total            9    0.288884
```

The constant term and all three coefficients of the regression equation give significant values of t (called 'T' in Minitab print-outs) suggesting that all three predictors, A_1, A_2 and A_3, should be included in the regression equation. The value of R^2 (adjusted) gives a measure of the predictive ability of the regression equation. Figure 8.11 shows a plot of the residuals against the fitted values: the residuals do not show any particular pattern. Figure 8.12 plots the predicted values against the measured values. The points are reasonably close to a straight line with no obvious outliers.

The corresponding equations for c_2, c_3 and c_4 are:

```
c2 = - 14.0 + 0.0179 A1 + 0.0821 A2 + 0.106  A3
c3 = - 9.84 + 0.0846 A1 - 0.0454 A2 + 0.0633 A3
c4 = - 4.51 + 0.0486 A1 + 0.112  A2 - 0.0834 A3
```

In Example 8.12.1 multiple regression was a suitable technique since there are only three predictor variables and, as the correlation matrix in Table 8.4 shows, the correlation between them is not very high.

In practice, a UV absorbance spectrum containing many hundreds of measurements would be used, rather than measurements of absorbance at just three wavelengths as in this example. As a result the number of possible predictor variables is likely to be much greater than the number of calibration specimens. Since it is not possible to carry out multiple regression in these circumstances the number of predictor variables must be reduced. One solution would be to use the intensity measurements at just some wavelengths but this poses the problem of deciding which would be the best wavelengths to choose. It would also mean that a large amount of data (and the information which they contain) would be discarded. There

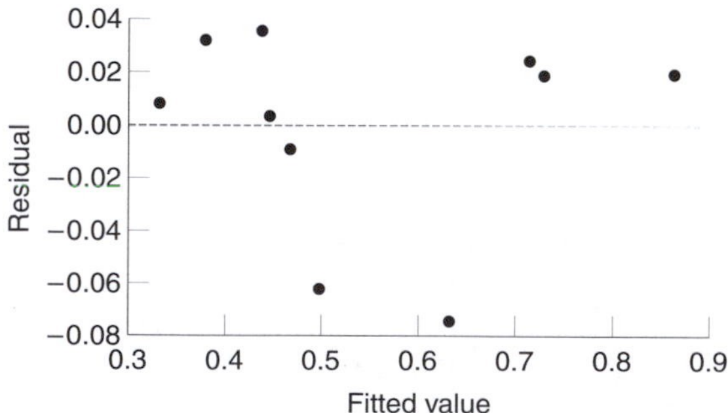

Figure 8.11 A plot of the residuals against the fitted values for Example 8.12.1.

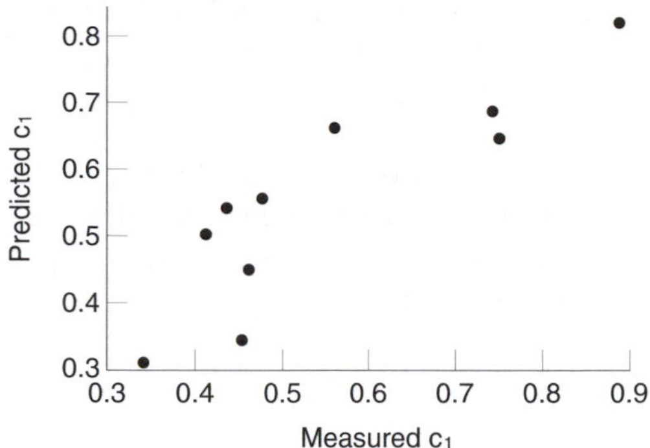

Figure 8.12 A plot of the predicted values of c_1 against the measured values for Example 8.12.1.

Table 8.4 Correlation matrix for the concentrations in Example 8.12.1

	c1	c2	c3
c2	-0.647		
c3	-0.706	0.094	
c4	-0.776	0.410	0.281

might also be a problem in using multiple regression because of correlation between the predictor variables. Both these problems can be overcome by using PCR or PLS1. These methods reduce the number of predictor variables to just a few and also give satisfactory results when there is correlation between the predictor variables. The preferred method in a given situation will depend on the precise nature of the data: an analysis can be carried out by each method and the results evaluated in order to find which method performs better.

Many recent applications of PCR and PLS have arisen in molecular spectroscopy, where strongly overlapping absorption and emission spectra often arise, even in simple mixtures. For example, a pesticide and its metabolites have been successfully analysed using Fourier transform infra-red spectroscopy and a mixture of very similar phenols was resolved by means of their fluorescence excitation spectra.

8.13 Artificial neural networks

No chapter on modern chemometric methods would be complete without a mention of **artificial neural networks** (ANN). In a simple form these attempt to imitate the operation of neurons in the brain. Such networks have a number of linked layers of artificial neurons, including an input and an output layer. Such a network could be used, for example, to classify specimens into one of several known classes. The network is trained by using a (large) training set. Its success at discrimination can be evaluated using a test set.

Neural networks find application in many other areas, for example pattern recognition and calibration. Many neural network designs have been studied but the circumstances, if any, in which they are superior to the other methods described in this chapter are not clear.

8.14 Conclusions

The aim of this chapter has been to give an introduction to the methods of multivariate analysis which are most commonly used in analytical chemistry. In most cases there is a choice of several different multivariate methods which could be applied to the same set of data. For example, in cluster analysis there is a choice between a hierarchical and a non-hierarchical approach, and each of these approaches offers a choice of several different methods. In multivariate calibration there is a choice between multiple regression, PCR and PLS regression. In addition, several approaches might be tried in the initial analysis. For example, cluster analysis and principal components analysis might be used prior to linear discriminant analysis, in order to see whether the objects being analysed fall naturally into groups.

There are many other methods than those described. Finally we should remember that multivariate analysis is a rapidly developing field, with new methods always becoming available as the power and speed of desktop computers grow.

Bibliography

Adams, M. J. 1995. *Chemometrics in Analytical Spectroscopy*. The Royal Society of Chemistry, Cambridge. (A tutorial guide to the application of the more commonly encountered techniques used in processing and interpreting analytical spectroscopic data.)

Chatfield, C. and Collins, A. J. 1980. *An Introduction to Multivariate Analysis*. Chapman & Hall, London. (Gives a general introduction to multivariate analysis, with a blend of theory and practice.)

Flury, B. and Riedwyl, H. 1988. *Multivariate Statistics – A Practical Approach*. Chapman & Hall, London. (Introduces selected methods of multivariate analysis at a non-technical level, with an emphasis on the basic principles underlying multivariate analysis.)

Manly, B. F. J. 1994. *Multivariate Statistical Methods – A Primer*. 2nd Edn. Chapman & Hall, London. (A general introduction to multivariate analysis at a non-technical level.)

Martens, H. and Naes, T. 1989. *Multivariate Calibration*. John Wiley and Sons Ltd, Chichester. (The book is structured so as to give a tutorial on the practical use of multivariate calibration techniques. It compares several calibration models, validation approaches and ways to optimize models.)

Otto, M. 1999. *Chemometrics*: *Statistics and Computer Application in Analytical Chemistry*. Wiley-VCH, Weinheim. (Gives a detailed treatment of the contents of this chapter.)

Vandeginste, B. G. M., Massart, D. L., Buydens, L. M. C., De Jong, S., Lewi, P. L. and Smeyers-Verbecke, J. 1998. *Handbook of Chemometrics and Qualimetrics: Part B*. Elsevier, Amsterdam. (A detailed and comprehensive account of the application of multivariate techniques in analytical chemistry.)

Exercises

1. For the data in Example 8.5.1 carry out a linear discriminant analysis working with the standardized variables. Hence identify the two variables which are most effective at discriminating between the two groups. Repeat the discriminant analysis with these two variables. Use the cross-classification success rate to compare the performance using two variables with that using all four variables.

2. The data below give the concentration (in mg kg^{-1}) of four elements found in samples of rice. The rice was one of two types: polished (P) or unpolished (U), was of one of two varieties (A or B) and was grown either in the wet season (W) or the dry season (D).

Variety	*Type*	*Season*	*P*	*K*	*Ni*	*Mo*
A	U	D	3555	2581	0.328	0.535
A	U	D	3535	2421	0.425	0.538
A	U	D	3294	2274	0.263	0.509
A	P	D	1682	1017	0.859	0.494
A	P	D	1593	1032	1.560	0.498
A	P	D	1554	984	1.013	0.478
B	U	D	3593	2791	0.301	0.771
B	U	D	3467	2833	0.384	0.407
B	P	D	2003	1690	0.216	0.728
B	P	D	1323	1327	0.924	0.393
A	U	W	3066	1961	0.256	0.481
A	P	W	1478	813	0.974	0.486
B	U	W	3629	2846	1.131	0.357
B	U	W	3256	2431	0.390	0.644
B	P	W	2041	1796	0.803	0.321
B	P	W	1745	1383	0.324	0.619

(Adapted from Phuong, T. D., Choung, P. V., Khiem, D. T. and Kokot, S. 1999. *Analyst* 124: 553)

(a) Carry out a cluster analysis. Do the samples appear to fall into groups? What characteristic is important in determining group membership?

(b) Calculate the correlation matrix. Which pairs of variables are strongly correlated? Which variable(s) show little correlation with the other variables?

(c) Carry out a principal components analysis and obtain a score plot. Does it confirm your analysis in (a)?

(d) Is it possible to identify the variety of a sample of rice by measuring the concentration of these four elements? Answer this question by carrying out a linear discriminant analysis. Investigate whether it is necessary to measure the concentration of all four elements in order to achieve satisfactory discrimination.

Solutions to exercises

(NB. Outline solutions are provided here: fuller solutions with commentaries are included in the Instructors' Manual.)

Chapter 1

1. Mean results ($g\ l^{-1}$) for laboratories A–E are: 41.9, 41.9, 43.2, 39.1, 41.5. Hence A – precise, little bias, mean accurate; B – poor precision, little bias, mean accurate but not very reliable; C – precise but biased to high values, poor accuracy; D – poor precision, biased to low values, poor accuracy; E – similar to A, but the last result might be an 'outlier'.
2. Laboratory A still shows little bias, but precision is poorer, reflecting reproducibility (i.e. between-day precision) rather than repeatability (within-day precision).
3. Number of binding sites must be an integer, clearly 2 here, so results are precise, but biased to low values. The bias does not matter much, as two binding sites can be deduced
4. (i) Blood lactate levels vary a lot in healthy patients, so great precision and accuracy not needed. (ii) Unbiased results could be crucial because of the great economic importance of U. (iii) Speed of analysis is crucial here, so precision and accuracy are less important. (iv) The aim is to detect even small changes over time, so precision is most important.
5. (i) Sample might not be representative, and/or reduction of Fe(III) to Fe(II) might be incomplete, giving biased results in each case. Completeness of reduction could be tested using a standard material. Random errors in each stage, including titrimetry, where they should be small. (ii) Sampling problem as in (i), and also incomplete extraction, leading

to bias (checked with standard). Random errors in spectrometry, which again should be relatively small. (iii) Random errors in gravimetry should be very small: more significant will be chemical problems such as co-precipitation, giving biased results.

Chapter 2

1. Mean = 0.077 μg ml^{-1}, s.d. = 0.007 μg ml^{-1}, RSD = 9%.
2. (i) 5.163 ± 0.025; (ii) 5.163 ± 0.038.
3. Mean = 22.3 ng ml^{-1}, s.d. = 1.4 ng ml^{-1}, RSD = 6.2%, 99% C.I. = 22.3 ± 1.4 ng ml^{-1}. Mean = 12.83 ng ml^{-1}, s.d. = 0.95 ng ml^{-1}, RSD = 7.4%, 99% C.I. = 12.8 ± 1.6 ng ml^{-1}.
4. 10.12 ± 0.18 ng ml^{-1}. Approximately 160.
5. 49.5 ± 1.1 ng ml^{-1}. Yes.
6. 10.18 ± 0.23 ml. No evidence for systematic error.
7. For weight of reagent: s.d. = 0.14 mg, RSD = 0.028% (0.029%).
   For volume of solvent: RSD = 0.02%.
   For molarity: RSD = 0.034% (0.020%).
   Values for reagent with formula weight 392 are given in brackets.
8. s.d. = 0.44×10^{-6} M.

Chapter 3

1. The points lie approximately on a straight line indicating that the data are drawn from a normal distribution.
2. t = 1.54, 1.60, 1.18, 1.60. None of means differs significantly from certified value.
3. (a) $Q = 0.565$ or $G = 1.97$. Not significant at $P = 0.05$. (b) $F = 34$. Significant at $P = 0.05$.
4. (a) $F = 1.70$. Not significant at $P = 0.05$. (b) $t = \pm 1.28$. Not significant at $P = 0.05$.
5. Between sample mean square = 2121.9, within sample mean square = 8.1. $F = 262$. Highly significant difference between depths. All pairs, except deepest pair, differ significantly from each other.
6. $t = \pm 1.20$. Sexes do not differ significantly.
7. $X^2 = 16.8$. No evidence that some digits are preferred to others.
8. Pine: $t = \pm 2.27$, not significant. Beech: $t = \pm 5.27$, significant at $P = 0.01$. Aquatic: $t = \pm 3.73$, significant at $P = 0.01$.

9. (a) $X^2 = 5.95$. The first worker differs significantly from the other three. (b) $X^2 = 2.81$. The last three workers do not differ significantly from each other.

10. $t = \pm 1.02$. Methods do not differ significantly.

11. Between-samples mean square = 0.1144, within samples mean square = 0.0445. $F = 2.57$. Just significant at $P = 0.05$. Least significant difference (0.25) indicates that A differs from B, D and E.

12. $t = \pm 2.2$. Men and women differ significantly.

13. $t = \pm 3.4$. Methods differ significantly.

14. Minimum size is 12.

Chapter 4

1. For scheme 1, $\sigma^2 = (4/2) + (10/5) = 4$. For scheme 2, $\sigma^2 = 4/(2 \times 3) + 10/3 = 4$. If S is the cost of sampling and A the cost of the analysis, then (cost of scheme 1/cost of scheme 2) = $(5S + 2A)/(3S + 6A)$. This ratio is >1 if $S/A > 2$.

2. ANOVA calculations show that the mean squares for the between-days and within-days variations are 111 and 3.25 respectively. Hence $F = 111/3.25 = 34$. The critical value of $F_{3,8}$ is 4.066 ($P = 0.05$), so the mean concentrations differ significantly. The sampling variance is given by $(111 - 3.25)/3 = 35.9$.

3. The mean squares for the between-sample and within-sample variations are 8.31×10^{-4} and 1.75×10^{-4} respectively, so $F = 8.31/1.75 = 4.746$. The critical value of $F_{3,8}$ is 4.066 ($P = 0.05$), so the between-sample mean square cannot be explained by measurement variation only. The latter variation, σ_0^2, is estimated as 1.75×10^{-4}. The estimate of the sampling variance, σ_1^2, is $([8.36 - 1.75] \times 10^{-4})/3 = 2.19 \times 10^{-4}$. Hence the variance of the mean for scheme 1 is $0.000175/4 + 0.000219/6 = 0.00008025$, and the variance of the mean for scheme 2 is $(0.000175/[2 \times 3]) + 0.000219/3 = 0.0001022$.

4. The six samples give six estimates of σ^2, which have an average of 2.795. So $\sigma = 1.67$. Hence the action and warning lines are at $50 \pm (2 \times 1.67)/\sqrt{4}$ and $50 \pm (3 \times 1.67)/\sqrt{4}$ respectively, i.e. at 50 ± 1.67 and 50 ± 2.50 respectively.

5. Samples A and B give mean values of 7.01 and 7.75 ppm respectively. Using a table of D and T values (e.g. for laboratory 1 these are −1.2 and 18.8 respectively), we find that $s_R^2 = 11.027$ and $s_r^2 = 0.793$. So $F = 11.027/0.793 = 13.905$, far higher than the critical $F_{14,14}$ value of ca. 2.48 ($P = 0.05$), obtained from the table by interpolation. Systematic errors are thus significant, and s_i^2 is found to be 5.117.

6. For the Shewhart chart for the mean, the values of W and A are found from tables ($n = 5$) to be 0.3768 and 0.5942 respectively. Hence the warning lines are at $120 \pm (7 \times 0.3768) = 120 \pm 2.64$, and the action lines are at $120 \pm (7 \times 0.5942) = 120 \pm 4.16$. For the range chart, the tables give w_1, w_2, a_1 and a_2 as 0.3653, 1.8045, 0.1580 and 2.3577 respectively so the lower warning line is at $7 \times 0.3653 = 2.56$, the upper warning line is at 12.63, and the lower and upper action lines are at 1.11 and 16.50 respectively.

7. Since $\sigma = 0.6$ and $n = 4$, the warning and action lines for the Shewhart chart for the mean are at 80 ± 0.6 and 80 ± 0.9 respectively. On this chart, the points for days 14–16 fall between the warning and action lines and point 17 is below the lower action line. So the chart suggests that the analytical process has gone out of control at about day 14. The cusum chart shows a steady negative trend from day 9 onwards, suggesting that the method was going out of control a good deal earlier.

Chapter 5

1. Here $r = -0.8569$. This suggests a strong correlation; Eq. (5.3) gives $t = 3.33$, well above the critical value ($P = 0.05$) of 2.78. But (a) a non-linear relationship is more likely, and (b) correlation is not causation – the Hg contamination may arise elsewhere.

2. In this case $r = 0.99982$. But the increase in the value of y (absorbance) with x is by a slightly decreasing amount at each point, i.e. this is really a curve, though little harm would come from treating it as a straight line.

3. The usual equations give $a = 0.0021$, $b = 0.0252$ and $s_{y/x} = 0.00703$. We then obtain $s_a = 0.00479$ and $s_b = 0.000266$. To convert the two latter values into 95% confidence intervals we multiply by $t = 2.57$, giving intervals for the intercept and slope of 0.0021 ± 0.0123 and 0.0252 ± 0.0007 respectively.

4. (a) A y-value of 0.456 corresponds to a concentration of 18.04 ng ml^{-1}. The s_{x_0} value is 0.300 so the confidence limits are $18.04 \pm (2.57 \times 0.300) = 18.04 \pm 0.77$ ng ml^{-1}. (b) The Q-test shows that the absorbance reading of 0.347 can be rejected as an outlier, the mean of the remaining three readings being 0.311, i.e. a concentration of 12.28 ng ml^{-1}. With $m = 3$ in this case, $s_{x_0} = 0.195$, giving confidence limits of 12.28 ± 0.50 ng ml^{-1}.

5. The absorbance at the limit of detection is given by $a + 3s_{y/x} = 0.0021 + (3 \times 0.00703) = 0.0232$. This corresponds to an x-value of 0.84 ng ml^{-1}, which is the limit of detection.

6. Here $a = 0.2569$ and $b = 0.005349$, so the Au concentration is $0.2569/0.005349 = 48.0$ ng ml^{-1}. The value of $s_{y/x}$ is 0.003693, so s_{x_E} is 0.9179. In this case $t = 2.45$, so the 95% confidence limits for the concentration are $48.0 \pm (2.45 \times 0.9179) = 48.0 \pm 2.2$ ng ml^{-1}.

7. The unweighted regression line has $b = 1.982$ and $a = 2.924$ respectively. Intensity values of 15 and 90 correspond to 6.09 and 43.9 ng ml^{-1} respectively. Then $s_{y/x} = 2.991$ and $s_{x_E} = 1.767$. So the confidence limits for the two concentrations are 6.09 ± 4.9 and 43.9 ± 4.9 ng ml^{-1}. The weighted line is found from the s values for each point, in increasing order 0.71, 0.84, 0.89, 1.64, 2.24, 3.03. The corresponding weights are 2.23, 1.59, 1.42, 0.42, 0.22 and 0.12 (totalling 6 as expected). The weighted line then has $b = 1.964$ and $a = 3.483$, so the intensity values of 15 and 90 correspond to concentrations of 5.87 and 44.1 ng ml^{-1} respectively. Estimated weights for these two points are 1.8 and 0.18 respectively, giving $s_{x_{0w}}$ values of 0.906 and 2.716, and confidence limits of 5.9 ± 2.5 and 44.1 ± 7.6 ng ml^{-1}.

8. If the ISE results are plotted as y and the gravimetric data are plotted as x the resulting line has $a = 4.48$ and $b = 0.963$. The r-value is 0.970. The confidence limits for a are 4.5 ± 20.1, which includes zero, and the limits for b are 0.96 ± 0.20, which includes 1, so there is no evidence of bias between the two methods.

9. Inspection suggests that the plot is linear up to A = 0.7–0.8. The line through all six points gives $r = 0.9936$, and residuals of −0.07, −0.02, +0.02, +0.06, +0.07, and −0.07. The trend suggests a curve. The SS for these values is 0.0191. If the last value is omitted, we find $r = 0.9972$, the residuals are −0.04, 0, +0.02, +0.04 and −0.02 (SS = 0.0040). Similar calculations show that the fifth point can be omitted also, at some cost in the range of the experiment.

10. The two straight line graphs are $y = 0.0014 + 0.0384x$, and $y = 0.1058 - 0.012x$. These intersect at an x-value of $(0.1058 - 0.0014)/(0.0384 - [-0.012]) = (0.1044/0.0504) = 2.07$, suggesting the formation of a 2 : 1 DPA : europium complex.

11. The best quadratic fit is $y = 0.0165 + 0.600x - 0.113x^2$. This gives $R^2 = 0.9991$ and $R'^2 = 0.9981$. The cubic fit is $y = -0.00552 + 0.764x - 0.383x^2 + 0.117x^3$. This gives $R^2 = 0.9999$ and $R'^2 = 0.9997$, so is a rather better fit.

12. For a straight line, a quadratic fit and a cubic fit, the R^2 values are 0.9238, 0.9786 and 0.9786 respectively, suggesting that a quadratic fit will be excellent. This is confirmed by the R'^2 values, which are 0.9085, 0.9679 and 0.9573 respectively, the quadratic fit giving the highest value.

Chapter 6

1. Mean = 9.96 ml, median = 9.90 ml. Q-test shows that the 10.20 value cannot quite be omitted ($P = 0.05$). If it is omitted, mean = 9.88, median = 9.89. The median is insensitive to outliers.

2. Sign test: compared with the median, the values give signs of − + 0 + − + + + +. So eight signs, of which six are positive. Probability

of this is 0.29, i.e. > 0.05, so null hypothesis is retained: median sulphur content could be 0.10%. In the signed rank test the zero is neglected, and the ranked differences are −0.01, 0.01, 0.01, −0.02, 0.02, 0.02, 0.04, 0.07. So signed ranks are −2, 2, 2, −5, 5, 5, 6, 7. Negative ranks total (−)7, but at $P = 0.05$, critical region is ≤ 3. So the null hypothesis is again retained.

3. (RID−EID) results give signs of + − + + + + + + 0+. So nine results, eight positive. $P = 0.04$ for this outcome, so the null hypothesis (that the methods give indistinguishable results) can be rejected. In the signed rank test, the negative ranks total (−)2.5, well below the critical level of 5, so again the null hypothesis can be rejected.

4. Arranging the results in order, the median is 23.5. So individual values have signs + + + − − − − − ++. This sequence has three runs, but for $M = N = 3$, the critical value is 3, so the null hypothesis of a random sequence must be retained.

5. Mann−Whitney U-test: 'beer' values are expected to be larger than 'lager' values. Number of lager values greater than the individual values = 4.5 (1 tie). Critical value for a one-sided test is 5, so we can just reject the null hypothesis ($P = 0.05$). Tukey's Quick Test: count is 5.5, just below the critical value of 6. So tests disagree: more data needed.

6. For instruments A−G student rankings are 3, 1, 5, 4, 7, 6, 2, and staff rankings are 5, 3, 6, 2, 4, 7, 1. So the d-values are −2, −2, −1, 2, 3, −1, 1, and the d^2 values are 4, 4, 1, 4, 9, 1, 1, totalling 24. Hence $r_s = 1 - [(6 \times 24)/(7 \times 48)] = 0.571$. For $n = 7$ the critical value at $P = 0.05$ is 0.786: no evidence of correlation between student and staff opinions.

7. If the x values are the distances and the y values the mercury levels, Theil's method gives $a = 2.575$, $b = -0.125$. (The least squares method gives $a = 2.573$, $b = -0.122$.)

8. To test $\mu = 1.0$ and $\sigma = 0.2$, we write $z = (x - 1.0)/0.2$. When the resulting z-values (1.5, 2.5, etc.) are compared with the cumulative distribution function for the normal distribution, the maximum difference is +0.335 at $z = 1.5$. The critical value is 0.409 (specified distribution, $P = 0.05$), so the null hypothesis is accepted. The appearance of the curves shows that $\mu = 1.0$ is roughly correct, but $\sigma = 0.2$ is much too low. The estimated mean and standard deviation from the data are 1.08 and 0.41 respectively. When the new z-values (0.54, 1.02, etc.) are plotted the maximum difference is only 0.11 at $z = 0.54$. The critical value is 0.262 so the null hypothesis can be retained: the data fit this normal distribution very well.

9. If the nickel levels are replaced by ranks (one tie occurs) the sums of the ranks for the three samples are 39, 52.5, and 79.5. (These add up to 171, as expected for 18 values, as $1/2 \times 18/19 = 171$.) The corresponding value of $X^2 = 4.97$, below the critical value of 5.99 ($P = 0.05$, 2 degrees of freedom) so the null hypothesis of no significant difference in the nickel levels in the oils must be retained.

1. This is two-way ANOVA without replication. The between-row (i.e. between-solution) mean square is 0.00370 (3 d.f.); the between-column (i.e. between-method) mean square is 0.00601 (2 d.f.); and the residual mean square is 0.00470 (6 d.f.). The between-solution mean square is less than the residual one, so is not significant. Comparison of the between-method and residual mean squares gives $F = 0.00601/0.00470 = 1.28$. The critical value of $F_{2,6}$ ($P = 0.05$) is 5.14, so the between-method variation is not significant.

2. Again, a two-way ANOVA experiment without replication. The between-soil, between-day and residual mean squares are respectively 4.67 (4 d.f.), 144.8 (2 d.f.) and 26.47 (8 d.f.). The between-soil mean square is less than the residual mean square, so there are no significant differences between soils. Comparing the between-day and residual mean squares gives $F = 144.8/26.47 = 5.47$. The critical value of $F_{2,8}$ is 4.46, so this source of variation is significant at $P = 0.05$. The actual probability (Excel) is 0.0318.

3. Another two-way ANOVA experiment without replication. (Replication would be needed to study possible interaction effects.) The between-compound, between-molar ratio and residual mean squares are respectively 4204 (3 d.f.), 584 (2 d.f.) and 706 (6 d.f.). Thus molar ratios have no significant effect. Comparing the between-compound and residual mean squares gives $F = 4204/706 = 5.95$. The critical value of $F_{3,6}$ is 4.76 ($P = 0.05$), so this variation is significant. (P is given by Excel as 0.0313.) Common sense should be applied to these and all other data – diphenylamine seems to behave differently from the other three compounds.

4. The single factor effects are A: −0.0215, C: 0.0005, T: −0.0265. The two-factor effects are AC: −0.0005, CT: 0.0025, AT: −0.0065. The three factor effect ACT is −0.0005.

5. This is a two-way ANOVA experiment with replication. The mean squares for between-row, between-column, interaction, and residual variations are respectively 2.53 (2 d.f.), 0.0939 (2 d.f.), 0.0256 (4 d.f.), and 0.0406 (9 d.f.). The interaction mean square is less than the residual mean square, so sample–laboratory interactions are not significant. Comparing the between-column (i.e. between-laboratory) and the residual mean squares gives $F = 0.0939/0.0406 = 2.31$. The critical value of $F_{2,9}$ is 4.256 ($P = 0.05$), so the between-laboratory variation is not significant.

6. (a) The Golden Ratio is used to determine the starting pHs as $5 + (4/1.618) = 7.47$, and $9 - (4/1.618) = 6.53$. (b) Using the Fibonacci approach to achieve a 40-fold reduction in the optimum range, we use the terms F_7 and F_9 (as F_9 is the first Fibonacci term above 40) to give the ratio 21/55. The starting pHs are then $5 + ([21 \times 4]/55) = 6.53$ and

$9 - ([21 \times 4]/55) = 7.47$. These values are the same as in (a), demonstrating that the Golden Ratio method is a limiting form of the Fibonacci search, giving the same outcomes when the degree of optimization sought by the latter method is high. (c) When six experiments are to be performed the Fibonacci method uses F_6 and F_4 to form the fraction 5/13, so the starting pHs are $5 + (20/13)$ and $9 - (20/13)$, i.e. 6.54 and 7.46 (similar values again). The degree of optimization is $1/F_6$, i.e. 1/13, so the optimum pH range will be defined within an envelope of $4/13 = 0.31$ pH units.

7. Vertex 1 should be rejected. The new vertex 8 should have coordinates 5.8, 9.4, 18.1, 9.2, 8.8 for factors A–E respectively, all values being given to one decimal place.

Chapter 8

1. The print-out below was obtained using Minitab.

```
Linear Discriminant Function for Group
                   A         B          C
Constant    -14.538    -2.439     -8.782
Sucrose      15.039    -3.697    -11.342
Glucose      -1.829     2.931     -1.102
Fructose     -9.612     0.363      9.249
Sorbitol     -2.191    -0.229      2.421
```

This suggests that sucrose and fructose may be the variables which are most effective at discriminating between varieties.

The cross-classification success rate with just these two variables is:

```
Summary of Classification with Cross-validation

Put into       ....True Group....
Group               A        B        C
A                   5        0        0
B                   0        5        1
C                   0        0        4
Total N             5        5        5
N Correct           5        5        4
Proportion      1.000    1.000    0.800

N = 15     N Correct = 14     Proportion Correct = 0.933
```

2. (a) A dendrogram shows two clear groups with group membership depending on whether the rice is polished or not.

(b)

	P	K	Ni
K	0.954		
Ni	-0.531	-0.528	
Mo	0.150	0.117	-0.527

Strong positive correlation between P and K. Little correlation between Mo and K and between Mo and P.

(c) Carrying out PCA on the standardized values gives:

Eigenanalysis of the Correlation Matrix

Eigenvalue	2.4884	1.1201	0.3464	0.0451
Proportion	0.622	0.280	0.087	0.011
Cumulative	0.622	0.902	0.989	1.000

Variable	PC1	PC2
P	0.577	0.340
K	0.572	0.366
Ni	-0.509	0.357
Mo	0.283	-0.789

A score plot shows two fairly well-defined groups: one for polished the other for unpolished samples.

(d) The results of LDA using the standardized values are:

Summary of Classification with Cross-validation

Put into	....True	Group....
Group	A	B
A	7	1
B	1	7
Total N	8	8
N Correct	7	7
Proportion	0.875	0.875

N = 16   N Correct = 14   Proportion Correct = 0.875

Linear Discriminant Function for Group

	A	B
Constant	-2.608	-2.608
P	18.016	-18.016
K	-19.319	19.319
Ni	-0.051	0.051
Mo	-1.198	1.198

The discrimination between varieties is good (87.5% success). Results suggest that P and K are most effective at discriminating between varieties. Using just these two elements a cross-classification rate of 15/16 is achieved.

APPENDIX 1

Commonly used statistical significance tests

Problem	*Tests available*	*See Section*	*Comments*
Testing for outliers	1. Dixon's test	3.7	
	2. Grubbs' tests	3.7	ISO recommended
Comparison of mean/ median with standard value	3. *t*-test	3.2	
	4. Sign test	6.3	Non-parametric
	5. Wilcoxon signed rank test	6.5	Non-parametric
Comparison of spreads of two data sets	6. *F*-test	3.6	Precedes test 8
	7. Siegel–Tukey test	6.6	Non-parametric
Comparison of means or medians of two samples	8. *t*-test	3.3	
	9. Mann–Whitney *U*-test	6.6	Non-parametric
	10. Tukey's quick test	6.6	Non-parametric
Comparison of two sets of paired data	11. Paired *t*-test	3.4	Small range of values
	12. Sign test	6.3	Non-parametric
	13. Wilcoxon signed rank test	6.5	Non-parametric
	14. x–y plot	5.9	Large range of values
Comparison of means/ medians of >2 samples	15. ANOVA	3.9	See index
	16. Kruskal–Wallis test	6.7	Non-parametric
Comparison of >2 matched data sets	17. Friedman's test	6.7	Non-parametric
Testing for occurrence of a particular distribution	18. Chi-squared test	3.11	
	19. Kolmogorov–Smirnov tests	6.12	Small samples

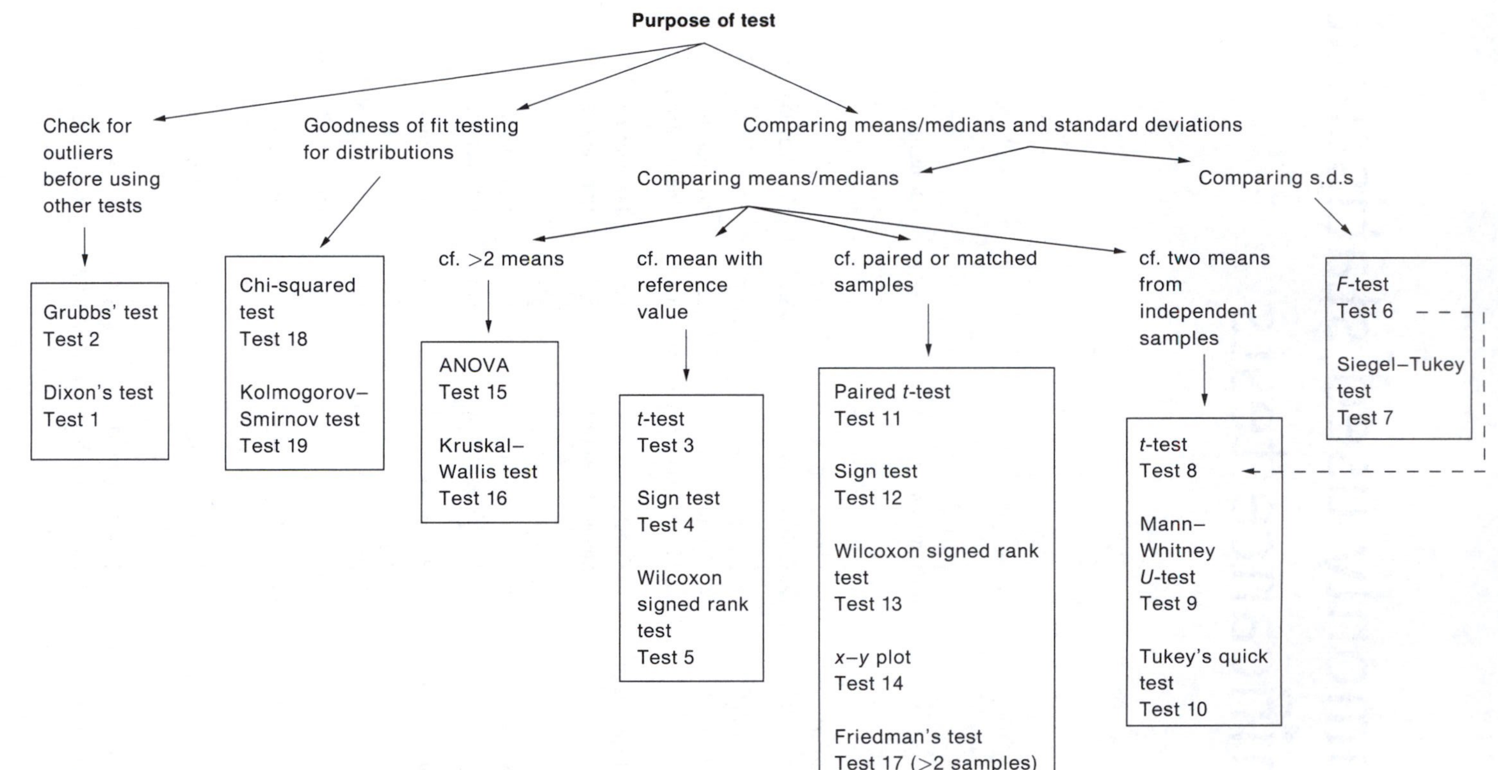

Flow chart for statistical significance tests

The flow chart

The flow chart is designed for use in conjunction with the table to aid the choice of appropriate significance test. It is intended only as a guide, and should not be used blindly. That is, once the chart has indicated which test or tests are most suitable to a given experimental situation, the analyst must become familiar with the principles of the selected test, the reasons for its selection, any limitations on its validity and so on. Only in this way will the results of the test be applied properly in all cases. For example, most non-parametric tests are not so powerful as parametric ones in conditions where the latter are appropriate, but may be more reliable where serious deviations from the normal distribution are known or suspected.

In the chart 'cf.' is used as an abbreviation for 'comparison of'. The test numbers refer to the table. Robust methods have not been included in either the table or the chart. Despite their growing importance they are still applied more usually by researchers and expert statisticians than by many laboratory workers, and the basic software packages referred to in Chapter 1 do not give a very comprehensive treatment of such methods. It is important to notice that ANOVA is a very widely used method, the exact form used depending on the problem to be solved: only the first reference to one-way ANOVA has been given in the table. The Cochran test (Section 4.11) and the least significant difference method (Section 3.9) used in conjunction with ANOVA, and the Wald–Wolfowitz test for runs (Section 6.4) have also been omitted for simplicity. The broken line linking Tests 6 and 8 is a reminder that, strictly speaking, the *F*-test should be applied to check whether the variances of the two samples under study are similar, before the *t*-test is applied. Some of the tests listed under 'Comparing means' actually compare medians; this has also been omitted in places in the interests of clarity.

Finally it is important to note that there are many tests in everyday use in addition to the ones listed above, as noted in the reference below.

Bibliography

Kanji, G. K. 1993. *100 Statistical Tests*. Sage Publications, London.

APPENDIX 2

Statistical tables

The following tables are presented for the convenience of the reader, and for use with the simple statistical tests, examples and exercises in this book. They are presented in a format that is compatible with the needs of analytical chemists: the significance level $P = 0.05$ has been used in most cases, and it has been assumed that the number of measurements available is fairly small. Most of these abbreviated tables have been taken, with permission, from *Elementary Statistics Tables* by Henry R. Neave, published by Routledge (Tables A.2–A.4, A.7, A.8, A.11–A.14). The reader requiring statistical data corresponding to significance levels and/or numbers of measurements not covered in the tables is referred to these sources.

Table A.1 $F(z)$, the standard normal cumulative distribution function

z	*0.00*	*0.01*	*0.02*	*0.03*	*0.04*	*0.05*	*0.06*	*0.07*	*0.08*	*0.09*
−3.4	0.0003	0.0003	0.0004	0.0004	0.0004	0.0004	0.0004	0.0004	0.0005	0.0005
−3.3	0.0005	0.0005	0.0005	0.0005	0.0006	0.0006	0.0006	0.0006	0.0006	0.0007
−3.2	0.0007	0.0007	0.0007	0.0008	0.0008	0.0008	0.0008	0.0009	0.0009	0.0009
−3.1	0.0010	0.0010	0.0010	0.0011	0.0011	0.0011	0.0012	0.0012	0.0013	0.0013
−3.0	0.0013	0.0014	0.0014	0.0015	0.0015	0.0016	0.0016	0.0017	0.0018	0.0018
−2.9	0.0019	0.0019	0.0020	0.0021	0.0021	0.0022	0.0023	0.0023	0.0024	0.0025
−2.8	0.0026	0.0026	0.0027	0.0028	0.0029	0.0030	0.0031	0.0032	0.0033	0.0034
−2.7	0.0035	0.0036	0.0037	0.0038	0.0039	0.0040	0.0041	0.0043	0.0044	0.0045
−2.6	0.0047	0.0048	0.0049	0.0051	0.0052	0.0054	0.0055	0.0057	0.0059	0.0060
−2.5	0.0062	0.0064	0.0066	0.0068	0.0069	0.0071	0.0073	0.0075	0.0078	0.0080
−2.4	0.0082	0.0084	0.0087	0.0089	0.0091	0.0094	0.0096	0.0099	0.0102	0.0104
−2.3	0.0107	0.0110	0.0113	0.0116	0.0119	0.0122	0.0125	0.0129	0.0132	0.0136
−2.2	0.0139	0.0143	0.0146	0.0150	0.0154	0.0158	0.0162	0.0166	0.0170	0.0174
−2.1	0.0179	0.0183	0.0188	0.0192	0.0197	0.0202	0.0207	0.0212	0.0217	0.0222
−2.0	0.0228	0.0233	0.0239	0.0244	0.0250	0.0256	0.0262	0.0268	0.0274	0.0281

Table A.1 Continued

z	*0.00*	*0.01*	*0.02*	*0.03*	*0.04*	*0.05*	*0.06*	*0.07*	*0.08*	*0.09*
−1.9	0.0287	0.0294	0.0301	0.0307	0.0314	0.0322	0.0329	0.0336	0.0344	0.0351
−1.8	0.0359	0.0367	0.0375	0.0384	0.0392	0.0401	0.0409	0.0418	0.0427	0.0436
−1.7	0.0446	0.0455	0.0465	0.0475	0.0485	0.0495	0.0505	0.0516	0.0526	0.0537
−1.6	0.0548	0.0559	0.0571	0.0582	0.0594	0.0606	0.0618	0.0630	0.0643	0.0655
−1.5	0.0668	0.0681	0.0694	0.0708	0.0721	0.0735	0.0749	0.0764	0.0778	0.0793
−1.4	0.0808	0.0823	0.0838	0.0853	0.0869	0.0885	0.0901	0.0918	0.0934	0.0951
−1.3	0.0968	0.0985	0.1003	0.1020	0.1038	0.1056	0.1075	0.1093	0.1112	0.1131
−1.2	0.1151	0.1170	0.1190	0.1210	0.1230	0.1251	0.1271	0.1292	0.1314	0.1335
−1.1	0.1357	0.1379	0.1401	0.1423	0.1446	0.1469	0.1492	0.1515	0.1539	0.1562
−1.0	0.1587	0.1611	0.1635	0.1660	0.1685	0.1711	0.1736	0.1762	0.1788	0.1814
−0.9	0.1841	0.1867	0.1894	0.1922	0.1949	0.1977	0.2005	0.2033	0.2061	0.2090
−0.8	0.2119	0.2148	0.2177	0.2206	0.2236	0.2266	0.2296	0.2327	0.2358	0.2389
−0.7	0.2420	0.2451	0.2483	0.2514	0.2546	0.2578	0.2611	0.2643	0.2676	0.2709
−0.6	0.2743	0.2776	0.2810	0.2843	0.2877	0.2912	0.2946	0.2981	0.3015	0.3050
−0.5	0.3085	0.3121	0.3156	0.3192	0.3228	0.3264	0.3300	0.3336	0.3372	0.3409
−0.4	0.3446	0.3483	0.3520	0.3557	0.3594	0.3632	0.3669	0.3707	0.3745	0.3783
−0.3	0.3821	0.3859	0.3897	0.3936	0.3974	0.4013	0.4052	0.4090	0.4129	0.4168
−0.2	0.4207	0.4247	0.4286	0.4325	0.4364	0.4404	0.4443	0.4483	0.4522	0.4562
−0.1	0.4602	0.4641	0.4681	0.4721	0.4761	0.4801	0.4840	0.4880	0.4920	0.4960
0.0	0.5000	0.5040	0.5080	0.5120	0.5160	0.5199	0.5239	0.5279	0.5319	0.5359
0.1	0.5398	0.5438	0.5478	0.5517	0.5557	0.5596	0.5636	0.5675	0.5714	0.5753
0.2	0.5793	0.5832	0.5871	0.5910	0.5948	0.5987	0.6026	0.6064	0.6103	0.6141
0.3	0.6179	0.6217	0.6255	0.6293	0.6331	0.6368	0.6406	0.6443	0.6480	0.6517
0.4	0.6554	0.6591	0.6628	0.6664	0.6700	0.6736	0.6772	0.6808	0.6844	0.6579
0.5	0.6915	0.6950	0.6965	0.7019	0.7054	0.7088	0.7123	0.7157	0.7190	0.7224
0.6	0.7257	0.7291	0.7324	0.7357	0.7389	0.7422	0.7454	0.7486	0.7517	0.7549
0.7	0.7580	0.7611	0.7642	0.7673	0.7704	0.7734	0.7764	0.7794	0.7823	0.7852
0.8	0.7881	0.7910	0.7939	0.7967	0.7995	0.8023	0.8051	0.8078	0.8106	0.8133
0.9	0.8159	0.8186	0.8212	0.8238	0.8264	0.8289	0.8315	0.8340	0.8365	0.8389
1.0	0.8413	0.8438	0.8461	0.8485	0.8508	0.8531	0.8554	0.8577	0.8599	0.8621
1.1	0.8643	0.8665	0.8686	0.8708	0.8729	0.8749	0.8770	0.8790	0.8810	0.8830
1.2	0.8849	0.8869	0.8888	0.8907	0.8925	0.8944	0.8962	0.8980	0.8997	0.9015
1.3	0.9032	0.9049	0.9066	0.9082	0.9099	0.9115	0.9131	0.9147	0.9162	0.9177
1.4	0.9192	0.9207	0.9222	0.9236	0.9251	0.9265	0.9279	0.9292	0.9306	0.9319
1.5	0.9332	0.9345	0.9357	0.9370	0.9382	0.9394	0.9406	0.9418	0.9429	0.9441
1.6	0.9452	0.9463	0.9474	0.9484	0.9495	0.9505	0.9515	0.9525	0.9535	0.9545
1.7	0.9554	0.9564	0.9573	0.9582	0.9591	0.9599	0.9608	0.9616	0.9625	0.9633
1.8	0.9641	0.9649	0.9656	0.9664	0.9671	0.9678	0.9686	0.9693	0.9699	0.9706
1.9	0.9713	0.9719	0.9726	0.9732	0.9738	0.9744	0.9750	0.9756	0.9761	0.9767
2.0	0.9772	0.9778	0.9783	0.9788	0.9793	0.9798	0.9803	0.9808	0.9812	0.9817

Table A.1 Continued

z	*0.00*	*0.01*	*0.02*	*0.03*	*0.04*	*0.05*	*0.06*	*0.07*	*0.08*	*0.09*
2.1	0.9821	0.9826	0.9830	0.9834	0.9838	0.9842	0.9846	0.9850	0.9854	0.9857
2.2	0.9861	0.9864	0.9868	0.9871	0.9875	0.9878	0.9881	0.9884	0.9887	0.9890
2.3	0.9893	0.9896	0.9898	0.9901	0.9904	0.9906	0.9909	0.9911	0.9913	0.9916
2.4	0.9918	0.9920	0.9922	0.9925	0.9927	0.9929	0.9931	0.9932	0.9934	0.9936
2.5	0.9938	0.9940	0.9941	0.9943	0.9945	0.9946	0.9948	0.9949	0.9951	0.9952
2.6	0.9953	0.9955	0.9956	0.9957	0.9959	0.9960	0.9961	0.9962	0.9963	0.9964
2.7	0.9965	0.9966	0.9967	0.9968	0.9969	0.9970	0.9971	0.9972	0.9973	0.9974
2.8	0.9974	0.9975	0.9976	0.9977	0.9977	0.9978	0.9979	0.9979	0.9980	0.9981
2.9	0.9981	0.9982	0.9982	0.9983	0.9984	0.9984	0.9985	0.9985	0.9986	0.9986
3.0	0.9987	0.9987	0.9987	0.9988	0.9988	0.9989	0.9989	0.9989	0.9990	0.9990
3.1	0.9990	0.9991	0.9991	0.9991	0.9992	0.9992	0.9992	0.9992	0.9993	0.9993
3.2	0.9993	0.9993	0.9994	0.9994	0.9994	0.9994	0.9994	0.9995	0.9995	0.9995
3.3	0.9995	0.9995	0.9995	0.9996	0.9996	0.9996	0.9996	0.9996	0.9996	0.9997
3.4	0.9997	0.9997	0.9997	0.9997	0.9997	0.9997	0.9997	0.9997	0.9997	0.9998

Table A.2 The *t*-distribution

Value of t for a confidence interval of *Critical value of $\|t\|$ for P values of number of degrees of freedom*	*90%* *0.10*	*95%* *0.05*	*98%* *0.02*	*99%* *0.01*
1	6.31	12.71	31.82	63.66
2	2.92	4.30	6.96	9.92
3	2.35	3.18	4.54	5.84
4	2.13	2.78	3.75	4.60
5	2.02	2.57	3.36	4.03
6	1.94	2.45	3.14	3.71
7	1.89	2.36	3.00	3.50
8	1.86	2.31	2.90	3.36
9	1.83	2.26	2.82	3.25
10	1.81	2.23	2.76	3.17
12	1.78	2.18	2.68	3.05
14	1.76	2.14	2.62	2.98
16	1.75	2.12	2.58	2.92
18	1.73	2.10	2.55	2.88
20	1.72	2.09	2.53	2.85
30	1.70	2.04	2.46	2.75
50	1.68	2.01	2.40	2.68
∞	1.64	1.96	2.33	2.58

The critical values of $|t|$ are appropriate for a *two*-tailed test. For a *one*-tailed test the value is taken from the column for *twice* the desired *P*-value, e.g. for a one-tailed test, $P = 0.05$, 5 degrees of freedom, the critical value is read from the $P = 0.10$ column and is equal to 2.02.

Table A.3 Critical values of F for a one-tailed test ($P = 0.05$)

v_2	v_1												
	1	*2*	*3*	*4*	*5*	*6*	*7*	*8*	*9*	*10*	*12*	*15*	*20*
1	161.4	199.5	215.7	224.6	230.2	234.0	236.8	238.9	240.5	241.9	243.9	245.9	248.0
2	18.51	19.00	19.16	19.25	19.30	19.33	19.35	19.37	19.38	19.40	19.41	19.43	19.45
3	10.13	9.552	9.277	9.117	9.013	8.941	8.887	8.845	8.812	8.786	8.745	8.703	8.660
4	7.709	6.944	6.591	6.388	6.256	6.163	6.094	6.041	5.999	5.964	5.912	5.858	5.803
5	6.608	5.786	5.409	5.192	5.050	4.950	4.876	4.818	4.772	4.735	4.678	4.619	4.558
6	5.987	5.143	4.757	4.534	4.387	4.284	4.207	4.147	4.099	4.060	4.000	3.938	3.874
7	5 591	4.737	4.347	4.120	3.972	3.866	3.787	3.726	3.677	3.637	3.575	3.511	3.445
8	5.318	4.459	4.066	3.838	3.687	3.581	3.500	3.438	3.388	3.347	3.284	3.218	3.150
9	5.117	4.256	3.863	3.633	3.482	3.374	3.293	3.230	3.179	3.137	3.073	3.006	2.936
10	4.965	4.103	3.708	3.478	3.326	3.217	3.135	3.072	3.020	2.978	2.913	2.845	2.774
11	4.844	3.982	3.587	3.357	3.204	3.095	3.012	2.948	2.896	2.854	2.788	2.719	2.646
12	4.747	3.885	3.490	3.259	3.106	2.996	2.913	2.849	2.796	2.753	2.687	2.617	2.544
13	4.667	3.806	3.411	3.179	3.025	2.915	2.832	2.767	2.714	2.671	2.604	2.533	2.459
14	4.600	3.739	3.344	3.112	2.958	2.848	2.764	2.699	2.646	2.602	2.534	2.463	2.388
15	4.543	3.682	3.287	3.056	2.901	2.790	2.707	2.641	2.588	2.544	2.475	2.403	2.328
16	4.494	3.634	3.239	3.007	2.852	2.741	2.657	2.591	2.538	2.494	2.425	2.352	2.276
17	4.451	3.592	3.197	2.965	2.810	2.699	2.614	2.548	2.494	2.450	2.381	2.308	2.230
18	4.414	3.555	3.160	2.928	2.773	2.661	2.577	2.510	2.456	2.412	2.342	2.269	2.191
19	4.381	3.522	3.127	2.895	2.740	2.628	2.544	2.477	2.423	2.378	2.308	2.234	2.155
20	4.351	3.493	3.098	2.866	2.711	2.599	2.514	2.447	2.393	2.348	2.278	2.203	2.124

v_1 = number of degrees of freedom of the numerator and v_2 = number of degrees of freedom of the denominator.

Table A.4 Critical values of F for a two-tailed test ($P = 0.05$)

v_2	v_1												
	1	*2*	*3*	*4*	*5*	*6*	*7*	*8*	*9*	*10*	*12*	*15*	*20*
1	647.8	799.5	864.2	899.6	921.8	937.1	948.2	956.7	963.3	968.6	976.7	984.9	993.1
2	38.51	39.00	39.17	39.25	39.30	39.33	39.36	39.37	39.39	39.40	39.41	39.43	39.45
3	17.44	16.04	15.44	15.10	14.88	14.73	14.62	14.54	14.47	14.42	14.34	14.25	14.17
4	12.22	10.65	9.979	9.605	9.364	9.197	9.074	8.980	8.905	8.844	8.751	8.657	8.560
5	10.01	8.434	7.764	7.388	7.146	6.978	6.853	6.757	6.681	6.619	6.525	6.428	6.329
6	8.813	7.260	6.599	6.227	5.988	5.820	5.695	5.600	5.523	5.461	5.366	5.269	5.168
7	8.073	6.542	5.890	5.523	5.285	5.119	4.995	4.899	4.823	4.761	4.666	4.568	4.467
8	7.571	6.059	5.416	5.053	4.817	4.652	4.529	4.433	4.357	4.295	4.200	4.101	3.999
9	7.209	5.715	5.078	4.718	4.484	4.320	4.197	4.102	4.026	3.964	3.868	3.769	3.667
10	6.937	5.456	4.826	4.468	4.236	4.072	3.950	3.855	3.779	3.717	3.621	3.522	3.419
11	6.724	5.256	4.630	4.275	4.044	3.881	3.759	3.664	3.588	3.526	3.430	3.330	3.226
12	6.554	5.096	4.474	4.121	3.891	3.728	3.607	3.512	3.436	3.374	3.277	3.177	3.073
13	6.414	4.965	4.347	3.996	3.767	3.604	3.483	3.388	3.312	3.250	3.153	3.053	2.948
14	6.298	4.857	4.242	3.892	3.663	3.501	3.380	3.285	3.209	3.147	3.050	2.949	2.844
15	6.200	4.765	4.153	3.804	3.576	3.415	3.293	3.199	3.123	3.060	2.963	2.862	2.756
16	6.115	4.687	4.077	3.729	3.502	3.341	3.219	3.125	3.049	2.986	2.889	2.788	2.681
17	6.042	4.619	4.011	3.665	3.438	3.277	3.156	3.061	2.985	2.922	2.825	2.723	2.616
18	5.978	4.560	3.954	3.608	3.382	3.221	3.100	3.005	2.929	2.866	2.769	2.667	2.559
19	5.922	4.508	3.903	3.559	3.333	3.172	3.051	2.956	2.880	2.817	2.720	2.617	2.509
20	5.871	4.461	3.859	3.515	3.289	3.128	3.007	2.913	2.837	2.774	2.676	2.573	2.464

v_1 = number of degrees of freedom of the numerator and v_2 = number of degrees of freedom of the denominator.

Table A.5 Critical values of Q ($P = 0.05$) for a two-sided test

Sample size	*Critical value*
4	0.831
5	0.717
6	0.621
7	0.570

Taken from King, E. P. 1958. *J. Am. Statist. Assoc.*, 48: 531.

Table A.6 Critical values of G ($P = 0.05$) for a two-sided test

Sample size	*Critical value*
3	1.155
4	1.481
5	1.715
6	1.887
7	2.020
8	2.126
9	2.215
10	2.290

Taken from *Outliers in Statistical Data*, Vic Barnett and Toby Lewis, 2nd Edition, 1984, John Wiley & Sons Limited.

Table A.7 Critical values of X^2 ($P = 0.05$)

Number of degrees of freedom	*Critical value*
1	3.84
2	5.99
3	7.81
4	9.49
5	11.07
6	12.59
7	14.07
8	15.51
9	16.92
10	18.31

Table A.8 Random numbers

02484	88139	31788	35873	63259	99886	20644	41853	41915	02944
83680	56131	12238	68291	95093	07362	74354	13071	77901	63058
37336	63266	18632	79781	09184	83909	77232	57571	25413	82680
04060	46030	23751	61880	40119	88098	75956	85250	05015	99184
62040	01812	46847	79352	42478	71784	65864	84904	48901	17115
96417	63336	88491	73259	21086	51932	32304	45021	61697	73953
42293	29755	24119	62125	33717	20284	55606	33308	51007	68272
31378	35714	00941	53042	99174	30596	67769	59343	53193	19203
27098	38959	49721	69341	40475	55998	87510	55523	15549	32402
66527	73898	66912	76300	52782	29356	35332	52387	29194	21591
61621	52967	40644	91293	80576	67485	88715	45293	59454	76218
18798	99633	32948	49802	40261	35555	76229	00486	64236	74782
36864	66460	87303	13788	04806	31140	75253	79692	47618	20024
10346	28822	51891	04097	98009	58042	67833	23539	37668	16324
20582	49576	91822	63807	99450	18240	70002	75386	26035	21459
12023	82328	54810	64766	58954	76201	78456	98467	34166	84186
48255	20815	51322	04936	33413	43128	21643	90674	98858	26060
92956	09401	58892	59686	10899	89780	57080	82799	70178	40399
87300	04729	57966	95672	49036	24993	69827	67637	09472	63356
69101	21192	00256	81645	48500	73237	95420	98974	36036	21781
22084	03117	96937	86176	80102	48211	61149	71246	19993	79708
28000	44301	40028	88132	07083	50818	09104	92449	27860	90196
41662	20930	32856	91566	64917	18709	79884	44742	18010	11599
91398	16841	51399	82654	00857	21068	94121	39197	27752	67308
46560	00597	84561	42334	06695	26306	16832	63140	13762	15598

Table A.9 The sign test

n	*r* = 0	*1*	*2*	*3*	*4*	*5*	*6*	*7*
4	0.063	0.313	0.688					
5	0.031	0.188	0.500					
6	0.016	0.109	0.344	0.656				
7	0.008	0.063	0.227	0.500				
8	0.004	0.035	0.144	0.363	0.637			
9	0.002	0.020	0.090	0.254	0.500			
10	0.001	0.011	0.055	0.172	0.377	0.623		
11	0.001	0.006	0.033	0.113	0.274	0.500		
12	0.000	0.003	0.019	0.073	0.194	0.387	0.613	
13	0.000	0.002	0.011	0.046	0.133	0.290	0.500	
14	0.000	0.001	0.006	0.029	0.090	0.212	0.395	0.605
15	0.000	0.000	0.004	0.018	0.059	0.151	0.304	0.500

The table uses the binomial distribution with $P = 0.5$ to give the probabilities of r or less successes for $n = 4$–15. These values correspond to a one-tailed sign test and should be doubled for a two-tailed test.

Table A.10 The Wald–Wolfowitz runs test

N	*M*	*At P = 0.05, the nurnber of runs is significant if it is:*	
		Less than	*Greater than*
2	12–20	3	NA
3	6–14	3	NA
3	15–20	4	NA
4	5–6	3	8
4	7	3	NA
4	8–15	4	NA
4	16–20	5	NA
5	5	3	9
5	6	4	9
5	7–8	4	10
5	9–10	4	NA
5	11–17	5	NA
6	6	4	10
6	7–8	4	11
6	9–12	5	12
6	13–18	6	NA
7	7	4	12
7	8	5	12
7	9	5	13
7	10–12	6	13
8	8	5	13
8	9	6	13
8	10–11	6	14
8	12–15	7	15

Adapted from Swed, F. S. and Eisenhart, C. 1943. *Ann. Math. Statist.*, 14: 66.
The test cannot be applied to data with *N*, *M* smaller than the given numbers, or to cases marked NA.

Table A.11 Wilcoxon signed rank test. Critical values for the test statistic at $P = 0.05$

n	*One-tailed test*	*Two-tailed test*
5	0	NA
6	2	0
7	3	2
8	5	3
9	8	5
10	10	8
11	13	10
12	17	13
13	21	17
14	25	21
15	30	25

The null hypothesis can be rejected when the test statistic is $\leq$ the tabulated value. NA indicates that the test cannot be applied.

Table A.12 Mann–Whitney *U*-test. Critical values for *U* or the lower of T_1 and T_2 at $P = 0.05$

n_1	n_2	*One-tailed test*	*Two-tailed test*
3	3	0	NA
3	4	0	NA
3	5	1	0
3	6	2	1
4	4	1	0
4	5	2	1
4	6	3	2
4	7	4	3
5	5	4	2
5	6	5	3
5	7	6	5
6	6	7	5
6	7	8	6
7	7	11	8

The null hypothesis can be rejected when *U* or the lower *T* value is $\leq$ the tabulated value. NA indicates that the test cannot be applied.

Table A.13 The Spearman rank correlation coefficient. Critical values for ρ at $P = 0.05$

n	*One-tailed test*	*Two-tailed test*
5	0.900	1.000
6	0.829	0.886
7	0.714	0.786
8	0.643	0.738
9	0.600	0.700
10	0.564	0.649
11	0.536	0.618
12	0.504	0.587
13	0.483	0.560
14	0.464	0.538
15	0.446	0.521
16	0.429	0.503
17	0.414	0.488
18	0.401	0.472
19	0.391	0.460
20	0.380	0.447

Table A.14 The Kolmogorov test. Critical two-tailed values for a specified distribution, and for unspecified normal distributions, at $P = 0.05$

n	*Specified distributions*	*Unspecified normal distributions*
3	0.708	0.376
4	0.624	0.375
5	0.563	0.343
6	0.519	0.323
7	0.483	0.304
8	0.454	0.288
9	0.430	0.274
10	0.409	0.262
11	0.391	0.251
12	0.375	0.242
13	0.361	0.234
14	0.349	0.226
15	0.338	0.219
16	0.327	0.213
17	0.318	0.207
18	0.309	0.202
19	0.301	0.197
20	0.294	0.192

The appropriate value is compared with the maximum difference between the experimental and theoretical cumulative frequency curves, as described in the text.

Table A.15 Critical values for C ($P = 0.05$) for $n = 2$

k	*Critical value*
3	0.967
4	0.906
5	0.841
6	0.781
7	0.727
8	0.680
9	0.638
10	0.602

Index